Physical Properties of Steroid Conjugates

by Seymour Bernstein, John P. Dusza, and Joseph P. Joseph

Lederle Laboratories

A Division of American Cyanamid Company

Pearl River, New York

SPRINGER SCIENCE+BUSINESS MEDIA, LLC 1968

ISBN 978-3-540-04060-6 ISBN 978-3-642-87828-2 (eBook)
DOI 10.1007/978-3-642-87828-2

Originally published by Springer-Verlag New York Inc. 1968
Softcover reprint of the hardcover 1st edition 1968
Library of Congress Catalog Card Number 68-9218
Title Number 1526

TO: *Ruth and Richard*

Acknowledgment

We wish to thank the following investigators for preprints of their forthcoming publications on conjugates: H. FEX (*Research Department, AB Leo*), and, D. FUKUSHIMA and G. SAUER (*Institute for Steroid Research, Montefiore Hospital Medical Center*).

The authors also wish to record here their deep appreciation and gratitude to Mrs. Dorothy Budd for her editorial assistance, and to Miss Elise Kramer for the typing of the manuscript.

Contents

Acknowledgment v

Introduction and Scope ix

Catalog of Conjugates 1

A. Empirical formula index of parent steroids 3

B. Catalog 12

References 199

Introduction and Scope

The initial purpose of this book was to provide a compilation of naturally occurring and synthetically prepared steroid conjugates and derivatives thereof with their physical constants as reported in the literature. A number of problems arose during the course of accumulating the necessary information. Foremost, it became evident that considerable characterization data were available which could not be strictly defined as "physical constants" according to organic chemical standards. Confronted with a mass of information of this type, and wishing to provide a broad general survey useful to many disciplines, the intent and scope of this book was then enlarged to encompass such data.

A general statement concerning the selection of compounds found in the book may be helpful to the reader. The selection was based on the following criteria: steroid glucuronides and sulfates isolated from natural sources or synthetically prepared for which (1) physical constant characterization data are provided, substantiating the structural assignment, or (2) chromatographic and/or allied data are presented establishing the compounds as glucuronides or sulfates. It is to be noted that mere characterization of a steroid obtained by chemical or enzymatic hydrolysis was generally not considered adequate qualification for inclusion of either the implicated or supposed glucuronide or sulfate into the compilation. However, certain compounds not conforming to the above selection guide lines, but deemed by us to be significant, were also included in this survey. It is then apparent that the compounds included in this catalog have been arbitrarily selected by the authors with primary emphasis being directed toward useful overall completeness rather than to rigorous selectivity of the data.

In order to assist in the retrieval of information from the catalog an explanation of the format for the presentation of the data is warranted. The catalog consists of one hundred and forty-seven entries, an entry being composed of all the conjugate derivatives of a parent steroid. The arrangement of entries is based on the empirical formula of the parent steroid and is designated on the page along with the steroid structure. In addition, the systematic nomenclature for the compound is provided along with the trivial name(s) associated with the compound. The above information is the standard heading for all entries regardless of the volume of the data included.

The catalog is primarily composed of derivatives of β-D-glucopyranosiduronic acid and sulfuric acid. Several other carbohydrate conjugates are also included. These are alphabetically arranged as major units in the catalog. Within each unit, position attachment to the steroid is indicated, and further arranged according to increasing carbon position number; the stereochemical configuration previously has been established in the depicted parent structure. Under each of these broad units are listed the various salts and derivatives for which data were obtained.

Citations of the individual compounds are arranged for an orderly presentation of the data. As a rule the molecular formula for the conjugate is given if it has been substantiated by elemental analytical results, these being indicated by the elements for which analytical data are presented in the research cited. The physical constants recorded are: melting point, optical rotation, ultraviolet absorption data, followed by supplementary spectral, chromatographic or allied information. Any part or all of the above characterization data are followed by a reference term given in brackets. The reference term is composed of a dual numbering system indicating both the year of the reference and an alphabetical listing within each year. Since an effort was made to collect all pertinent characterization data, some compounds have several references citing similar physical constants. In the case of frequently described conjugates it was deemed advisable to limit the number of citations. Successive entries for a compound appear without restatement of the empirical formula; however, the extent of their analytical characterization is always indicated. Solvates (hydrates) are treated as separate entries since certain of their physical constants may differ appreciably from those of the unsolvated compound. Solvent data associated with physical constant measurements are included in most cases. Solvents or solvent systems employed for recrystallization or chromatographic purposes are not included in the catalog. This information is obtainable from the original publications.

Some distinction was required in the sections of the catalog reporting physical constants of specific salts and derivatives from that part devoted to the supplementary characterization data of naturally occurring glucuronides and sulfates. Since for the latter no specific salts or derivatives are generally designated, the characterization data are confined to chromatographic or allied information. In those entries where both types of information were available, the supplementary data are grouped under the term *Allied Characterization Data* located after the specific data in each major subdivision. Where only supplementary data were available. the heading *Allied Characterization Data* is omitted.

The citations of patents from countries other than the U.S.A. after 1962 have also been given a Derwent Pharmaceutical Patents Documentation (FARMDOC) number for the corresponding basic patent to provide additional information. Prior to 1963, Chemical Abstracts reference are cited.

The following abbreviations are employed in this chapter, and their equivalence and scope defined:

CC = Column chromatography (includes absorption, partition, gel filtration, and ion exchange chromatography)

CCD = Countercurrent distribution

GLC = Gas-liquid chromatography

IR = Infrared spectrum (depicted in original text, or major bands cited)

MS = Mass spectrum

NMR = Nuclear magnetic resonance (usually major peaks cited)

ORD = Optical rotatory dispersion

PC = Paper chromatography (includes normal and impregnated papers)

PE = Paper electrophoresis

TLC = Thin layer chromatography (includes all the various coatings and techniques available).

This catalog was compiled from a survey of the literature as of December 1967, and represents a reasonably accurate coverage of the field. The authors would appreciate having any important omissions brought to their attention.

Catalog of Conjugates

A. Empirical formula index of parent steroids

$C_{18}H_{18}O_2$ 3-Hydroxyestra-1,3,5(10),6,8-pentaen-17-one (Equilenin) 12

$C_{18}H_{20}O_2$ Estra-1,3,5(10),6,8-pentaene-3,17β-diol (Dihydroequilenin) 13

3-Hydroxyestra-1,3,5(10),7-tetraen-17-one (Equilin) 14

$C_{18}H_{22}O_2$ Estra-1,3,5(10),6-tetraene-3,17β-diol (6-Dehydro-17β-estradiol) 15

Estra-1,3,5(10),7-tetraene-3,17β-diol (Dihydroequilin) 16

3-Hydroxyestra-1,3,5(10)-trien-17-one (Estrone) 17

$C_{18}H_{22}O_3$ 2,3-Dihydroxyestra-1,3,5(10)-trien-17-one (2-Hydroxyestrone) 20

3,16α-Dihydroxyestra-1,3,5,(10)-trien-17-one (16α-Hydroxyestrone) 21

3,16β-Dihydroxyestra-1,3,5(10)-trien-17-one (16β-Hydroxyestrone) 23

3,17β-Dihydroxyestra-1,3,5(10)-trien-16-one (16-Keto-17β-estradiol) 24

$C_{18}H_{24}O$ Estra-1,3,5(10)-trien-3-ol 25

$C_{18}H_{24}O_2$ Estra-1,3,5(10)-triene-3,17α-diol (17α-Estradiol) 26

Estra-1,3,5(10)-triene-3,17β-diol (17β-Estradiol, estradiol) 27

$C_{18}H_{24}O_3$ Estra-1,3,5(10)-triene-2,3,17β-triol (2-Hydroxy-17β-estradiol, 2-hydroxyestradiol) 33

Estra-1,3,5(10)-triene-3,15α,17β-triol (15α-Hydroxy-17β-estradiol, 15α-hydroxy-estradiol) 34

Estra-1,3,5(10)-triene-3,16α,17α-triol (17-Epiestriol) 35

Estra-1,3,5(10)-triene-3,16α,17β-triol (Estriol) 36

Estra-1,3,5(10)-triene-3,16β,17β-triol (16-Epiestriol) 44

$C_{18}H_{26}O_2$ 17β-Hydroxyestra-4-en-3-one (19-Nortestosterone) 46

$C_{19}H_{24}O_3$ Androst-4-ene-3,11,17-trione (Adrenosterone) 47

$C_{19}H_{26}O_2$ Androst-4-ene-3,17-dione (Androstenedione) 48

17β-Hydroxyandrosta-4,9(11)-dien-3-one (9-Dehydrotestosterone) 49

$C_{19}H_{26}O_3$ 11β-Hydroxyandrost-4-ene-3,17-dione 50

17β-Hydroxyandrost-4-ene-3,11-dione (11-Ketotestosterone) 51

3β-Hydroxyandrost-5-ene-7,17-dione (7-Ketodehydroisoandrosterone, 7-keto-dehydroepiandrosterone) 52

$C_{19}H_{28}O_2$ 5α-Androstane-3,17-dione (Androstanedione) 53

5β-Androstane-3,17-dione (Etiocholanedione) 54

17α-Hydroxyandrost-4-en-3-one (Epitestosterone) 55

17β-Hydroxyandrost-4-en-3-one (Testosterone) 56

$C_{19}H_{28}O_2$ (cont.)

3β-Hydroxyandrost-5-en-17-one (Dehydroisoandrosterone, dehydroepiandrosterone, androstenolone). 59

$C_{19}H_{28}O_3$

3α-Hydroxy-5α-androstane-11,17-dione (11-Ketoandrosterone) 63

3α-Hydroxy-5β-androstane-11,17-dione (11-Ketoetiocholanolone). 64

6β,17β-Dihydroxyandrost-4-en-3-one (6β-Hydroxytestosterone) 65

11α,17β-Dihydroxyandrost-4-en-3-one (11α-Hydroxytestosterone) 66

11β,17β-Dihydroxyandrost-4-en-3-one (11β-Hydroxytestosterone) 67

14α,17β-Dihydroxyandrost-4-en-3-one (14α-Hydroxytestosterone) 68

3β,7α-Dihydroxyandrost-5-en-17-one (7α-Hydroxydehydroisoandrosterone, 7α-hydroxydehydroepiandrosterone). 69

3β,16α-Dihydroxyandrost-5-en-17-one (16α-Hydroxydehydroisoandrosterone, 16α-hydroxydehydroepiandrosterone) 70

$C_{19}H_{30}O$

5α-Androst-16-en-3α-ol. 71

$C_{19}H_{30}O_2$

Androst-5-ene-3β,17α-diol 72

Androst-5-ene-3β,17β-diol (Androstenediol) 73

3α-Hydroxy-5α-androstan-17-one (Androsterone) 75

3β-Hydroxy-5α-androstan-17-one (Isoandrosterone, epiandrosterone) 78

3α-Hydroxy-5β-androstan-17-one (Etiocholanolone) 80

3β-Hydroxy-5β-androstan-17-one 82

17β-Hydroxy-5α-androstan-3-one 83

$C_{19}H_{30}O_3$	Androst-5-ene-3β,16α,17β-triol (Androstenetriol)	84
	Androst-5-ene-3β,16β,17α-triol	85
	3α-Hydroxy-17a-oxa-D-homo-5β-androstan-17-one	86
	3α,11β-Dihydroxy-5α-androstan-17-one (11β-Hydroxyandrosterone)	87
	3α,11β-Dihydroxy-5β-androstan-17-one (11β-Hydroxyetiocholanolone)	88
$C_{19}H_{32}O_2$	5α-Androstane-3α,17α-diol	89
	5α-Androstane-3α,17β-diol	90
	5α-Androstane-3β,17β-diol	91
	5β-Androstane-3α,17β-diol	92
	5β-Androstane-3β,17β-diol	93
$C_{19}H_{32}O_3$	5α-Androstane-3α,16β,17α-triol	94
$C_{20}H_{24}O_2$	17α-Ethynylestra-1,3,5(10)-triene-3,17β-diol (Ethynylestradiol)	95
$C_{20}H_{28}O_3$	17β-Carboxyandrost-4-en-3-one (3-Keto-4-etienic acid)	96
	17α-Hydroxy-19-norpregn-4-ene-3,20-dione	97
$C_{20}H_{30}O_2$	17β-Hydroxy-17α-methylandrost-4-en-3-one (Methyltestosterone)	98
$C_{20}H_{32}O_2$	17α-Methylandrost-5-ene-3β,17β-diol (Methylandrostenediol)	99
$C_{21}H_{26}O_3$	17α-Hydroxypregna-1,4,6-triene-3,20-dione	100
$C_{21}H_{26}O_5$	17α,21-Dihydroxypregna-1,4-diene-3,11,20-trione (Prednisone)	101
$C_{21}H_{27}FO_6$	9α-Fluoro-11β,16α,17α,21-tetrahydroxypregna-1,4-diene-3,20-dione (Triamcinolone)	102

$C_{21}H_{28}O_4$ 21-Hydroxypregn-4-ene-3,11,20-trione (Dehydrocorticosterone) 103

$C_{21}H_{28}O_5$ 11β,21-Dihydroxy-3,20-dioxopregn-4-en-18-al (Aldosterone) 104

17α,21-Dihydroxypregn-4-ene-3,11,20-trione (Cortisone) 105

11β,17α,21-Trihydroxypregna-1,4-diene-3,20-dione (Prednisolone) 107

$C_{21}H_{28}O_6$ 11β,16α,17α,21-Tetrahydroxypregna-1,4-diene-3,20-dione (16α-Hydroxyprednisolone) 109

$C_{21}H_{29}FO_5$ 9α-Fluoro-11β,17α,21-trihydroxypregn-4-ene-3,20-dione (9α-Fluorohydrocortisone, 9α-fluorocortisol) . . 110

$C_{21}H_{30}O_2$ 3β-Hydroxypregna-5,16-dien-20-one 111

Pregn-4-ene-3,20-dione (Progesterone) 112

$C_{21}H_{30}O_3$ 6β-Hydroxypregn-4-ene-3,20-dione (6β-Hydroxyprogesterone) 113

17α-Hydroxypregn-4-ene-3,20-dione (17α-Hydroxyprogesterone) 114

21-Hydroxypregn-4-ene-3,20-dione (Deoxycorticosterone, cortexone) 115

$C_{21}H_{30}O_4$ 11β,21-Dihydroxypregn-4-ene-3,20-dione (Corticosterone) 116

17α,21-Dihydroxypregn-4-ene-3,20-dione (Compound S) 117

$C_{21}H_{30}O_5$ 6β,11β,21-Trihydroxypregn-4-ene-3,20-dione (6β-Hydroxycorticosterone) 119

11β,17α,21-Trihydroxypregn-4-ene-3,20-dione (Hydrocortisone, cortisol) 120

$C_{21}H_{30}O_6$ 6β,11β,17α,21-Tetrahydroxypregn-4-ene-3,20-dione (6β-Hydroxyhydrocortisone, 6β-hydroxycortisol). 122

$C_{21}H_{32}O_2$ 3β-Hydroxypregn-5-en-20-one (Pregnenolone) 123

3β-Hydroxy-5α-pregn-16-en-20-one 125

$C_{21}H_{32}O_3$ 3α-Hydroxy-5β-pregnane-11,20-dione 126

3β,17α-Dihydroxypregn-5-en-20-one (17α-Hydroxypregnenolone) 127

3β,21-Dihydroxypregn-5-en-20-one 129

$C_{21}H_{32}O_4$ 3α,21-Dihydroxy-5α-pregnane-11,20-dione. . . 130

3α,21-Dihydroxy-5β-pregnane-11,20-dione. . . 131

3β,17α,21-Trihydroxypregn-5-en-20-one 132

$C_{21}H_{32}O_5$ 3α,11β,21-Trihydroxy-20-oxo-5β-pregnan-18-al (Tetrahydroaldosterone) 133

3α,17α,21-Trihydroxy-5β-pregnane-11,20-dione (Tetrahydro-E, tetrahydrocortisone). . . . 134

$C_{21}H_{34}O_2$ 3α-Hydroxy-5α-pregnan-20-one 135

3β-Hydroxy-5α-pregnan-20-one 136

3α-Hydroxy-5β-pregnan-20-one (Pregnanolone) 138

3β-Hydroxy-5β-pregnan-20-one 140

Pregn-5-ene-3β,20α-diol 141

Pregn-5-ene-3β,20β-diol 143

$C_{21}H_{34}O_3$ 3α,17α-Dihydroxy-5β-pregnan-20-one 144

3α,21-Dihydroxy-5β-pregnan-20-one. 145

Pregn-5-ene-3β,17α,20α-triol 146

$C_{21}H_{34}O_4$ 3α,11β,21-Trihydroxy-5α-pregnan-20-one . . . 147

3α,11β,21-Trihydroxy-5β-pregnan-20-one (Tetrahydrocorticosterone) 148

$C_{21}H_{34}O_4$ (cont.)	3β,11β,21-Trihydroxy-5α-pregnan-20-one	149
	3α,17α,21-Trihydroxy-5β-pregnan-20-one (Tetrahydro-S)	150
	3β,17α,21-Trihydroxy-5α-pregnan-20-one	151
$C_{21}H_{34}O_5$	3α,11β,17α,21-Tetrahydroxy-5β-pregnan-20-one (Tetrahydro-F, tetrahydrohydrocortisone, tetrahydrocortisol)	152
	3α,17α,20α,21-Tetrahydroxy-5β-pregnan-11-one (Cortolone)	154
	3α, 17α, 20β,21-Tetrahydroxy-5β-pregnan-11-one (β-Cortolone)	155
$C_{21}H_{36}O_2$	5α-Pregnane-3α,20α-diol	156
	5α-Pregnane-3β,20β-diol	157
	5β-Pregnane-3α,20α-diol (Pregnanediol)	158
	5β-Pregnane-3α,20β-diol	160
$C_{21}H_{36}O_3$	5α-Pregnane-3β,17α,20β-triol	161
	5β-Pregnane-3α,17α,20ξ triol	162
	5α-Pregnane-3β,20ξ,21-triol	163
$C_{21}H_{36}O_5$	5β-Pregnane-3α,11β,17α,20α,21-pentol (Cortol)	164
	5β-Pregnane-3α,11β,17α,20β,21-pentol (β-Cortol)	165
$C_{21}H_{37}NO$	20α-Amino-5α-pregnan-3β-ol	166
$C_{22}H_{28}O_3$	17α-Hydroxy-1α,2α-methanopregna-4,6-diene-3,20-dione	167
$C_{22}H_{29}FO_5$	9α-Fluoro-11β,17α,21-trihydroxy-16α-methyl-pregna-1,4-diene-3,20-dione (Dexamethasone)	168
	9α-Fluoro-11β,17α,21-trihydroxy-16β-methyl-pregna-1,4-diene-3,20-dione (Betamethasone)	169

$C_{22}H_{32}O_3$ 17α-Hydroxy-6α-methylpregn-4-ene-3,20-dione 170

$C_{22}H_{32}O_5$ 11β,17α,21-Trihydroxy-16α-methylpregn-4-ene-3,20-dione. 171

$C_{22}H_{34}O_2$ 3β-Hydroxy-17α-methylpregn-5-en-20-one . . 172

$C_{22}H_{34}O_3$ 3β,17α-Dihydroxy-6-methylpregn-5-en-20-one. 173

$C_{23}H_{34}O_4$ 3β,14β,Dihydroxy-5β-card-20(22)-enolide (Digitoxigenin) 174

$C_{27}H_{44}O$ Cholest-4-en-3-one (Cholestenone) 175

Cholesta-5,7-dien-3β-ol (7-Dehydrocholesterol) 176

Vitamin D_3 177

$C_{27}H_{46}O$ Cholest-5-en-3α-ol (Epicholesterol) 178

Cholest-5-en-3β-ol (Cholesterol) 179

$C_{27}H_{46}O_2$ Cholest-5-ene-3β,20α-diol (20α-Hydroxycholesterol) 182

$C_{27}H_{48}O$ 5α-Cholestan-3α-ol 183

5α-Cholestan-3β-ol (Cholestanol) 184

$C_{27}H_{48}O_2$ 5α-Cholestan-3β,7α-diol. 185

5α-Cholestane-3β,7β-diol 186

$C_{27}H_{48}O_3$ Cholestane-3β,5α,6β-triol. 187

$C_{27}H_{48}O_4$ 5α-Cholestane-3β,7α,16α,26-tetrol (Myxinol). 188

$C_{28}H_{44}O$	Ergosta-5,7,22-trien-3β-ol (Ergosterol)	189
	Vitamin D_2	190
$C_{29}H_{48}O$	Stigmasta-5,22-dien-3β-ol (Stigmasterol)	191
$C_{29}H_{50}O$	Stigmast-5-en-3β-ol (β-Sitosterol)	192
$C_{29}H_{52}O$	5α-Stigmastan-3β-ol	194
$C_{30}H_{50}O$	Lanosta-7,9(11)-dien-3β-ol	195
	Lanosta-8,24-dien-3β-ol (Lanosterol)	196
$C_{30}H_{52}O$	Lanost-8-en-3β-ol	197
$C_{31}H_{48}O_4$	Fusidic acid	198

B. Catalog

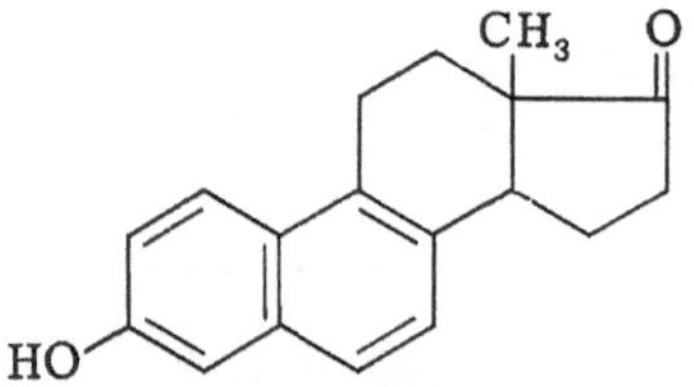

3-Hydroxyestra-1,3,5(10),6,8-pentaen-17-one
(*Equilenin*)

3-SULFATE

Quinidine salt – $C_{38}H_{42}N_2O_7S$ ⟨49-1⟩.

Sodium salt – $C_{18}H_{17}NaO_5S{\cdot}H_2O$ (anal. C, H, S), $[\alpha]_D^{20}$ + 70° (water) ⟨49-1⟩ • (anal. H_2O, sulfated ash) ⟨53-2⟩ • ^{35}S ⟨50-2⟩ • CC ⟨67-15⟩.

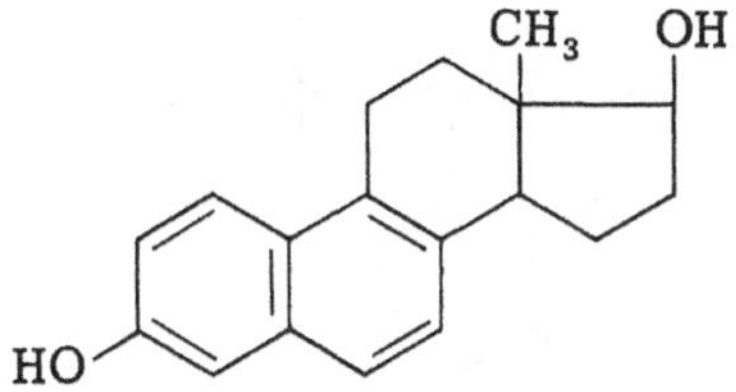

Estra-1,3,5(10),6,8-pentaene-3,17β-diol
(Dihydroequilenin)

3-SULFATE

CC ⟨67-15⟩.

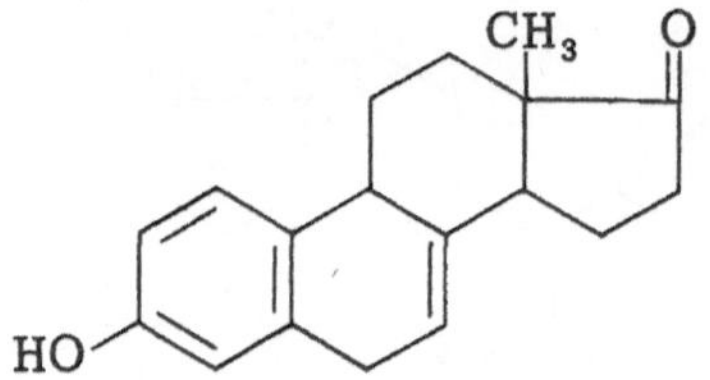

3-Hydroxyestra-1,3,5(10),7-tetraen-17-one
(Equilin)

3-SULFATE

1-Phenyl-2-aminopropane salt — $C_{27}H_{35}NO_5S$ (anal. N, S), mp 80-95° ⟨50-3⟩.

Potassium salt — $C_{18}H_{19}KO_5S \cdot 3H_2O$ (anal. S), $[\alpha]_D$ + 208° (water) ⟨52-2⟩.

Quinidine salt — $C_{38}H_{44}N_2O_7S$ (anal. N), $[\alpha]_D$ + 247° (methanol) ⟨52-2⟩ • $C_{38}H_{44}N_2O_7S$ (anal. N) ⟨49-1⟩.

Sodium salt — $C_{18}H_{19}NaO_5S$ (anal. C, H, S), mp 187-192°, $[\alpha]_D$ + 217° (water) ⟨52-2⟩ • (anal. C, H, S), $[\alpha]_D^{20}$ + 218° (water) ⟨49-1⟩ • (anal. H_2O, sulfated ash), $[\alpha]_D^{25}$ + 187.5° (methanol) ⟨53-2⟩ • CC ⟨67-15⟩ • ^{35}S ⟨50-2⟩.

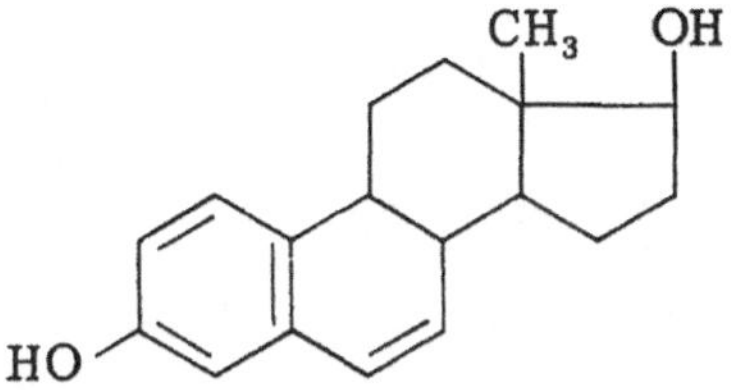

CH3 OH

HO

Estra-1,3,5(10),6-tetraene-3,17β-diol
(6-Dehydro-17β-estradiol)

17-β-D-GLUCOPYRANOSIDURONIC ACID

Sodium Salt – $C_{24}H_{29}NaO_8$ (anal. C, H), mp 230-240°, IR, TLC ⟨66-17⟩.

Estra-1,3,5(10),7-tetraene-3,17β-diol
(Dihydroequilin)

3-SULFATE

CC ⟨67-15⟩.

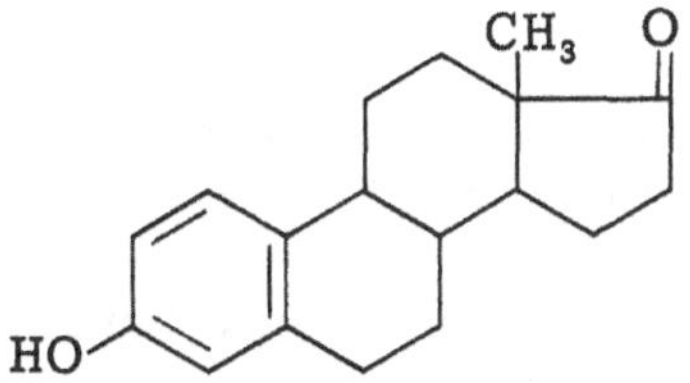

3-Hydroxyestra-1,3,5(10)-trien-17-one
(Estrone)

3-β-D-GLUCOPYRANOSIDURONIC ACID

Acid – $C_{24}H_{30}O_8$, mp 176.5-178° ⟨67-10⟩ • preparation but no data ⟨60-16⟩ • 6,7-^{3}H,G-^{14}C, biochemical preparation ⟨67-35⟩.

Sodium salt – $C_{24}H_{29}NaO_8 \cdot H_2O$ (anal. C, H), darkens 294°, mp > 310° ⟨67-10⟩.

2′,3′,4′-Tri-*O*-acetyl methyl ester – $C_{31}H_{38}O_{11}$, mp 225.5-228°, $[\alpha]_D^{25}$ + 57.1° (chloroform) ⟨39-3⟩ • $C_{31}H_{38}O_{11} \cdot 2H_2O$ (anal. C, H), mp 212-215°, $[\alpha]_D^{23}$ + 54° (chloroform) ⟨59-7⟩ • $C_{31}H_{38}O_{11}$ (anal. C, H), mp 226.5-227.5°, $[\alpha]_D$ + 57° (chloroform), IR ⟨67-10⟩ • IR ⟨59-8⟩ • preparation but no data ⟨60-16⟩ • 6,7-^{3}H ⟨67-35⟩.

Allied Characterization Data

3-GLUCURONIDE

CC ⟨66-23⟩⟨67-4⟩⟨67-13⟩⟨67-21⟩⟨67-29⟩ • CCD ⟨60-16⟩⟨62-4⟩⟨67-35⟩ • PC ⟨66-12⟩⟨66-23⟩⟨67-29⟩ • TLC ⟨62-4⟩⟨66-33⟩⟨67-25⟩⟨67-29⟩⟨67-35⟩.

3-GLUCURONIDE TRI-*O*-ACETYL METHYL ESTER

CCD ⟨62-4⟩⟨67-35⟩ • TLC ⟨62-4⟩.

* * * * *

3-SULFATE

Ammonium salt – $C_{18}H_{25}NO_5S \cdot 2H_2O$ (anal. H_2O, sulfated ash), $[\alpha]_D^{25}$ + 134° (water) ⟨53-2⟩.

Calcium salt – $(C_{18}H_{21}O_5S)_2Ca \cdot 3H_2O$ (anal. H_2O), $[\alpha]_D^{25}$ + 120° (water) ⟨53-2⟩.

Choline salt – mp 255-258° ⟨60-7⟩.

N,N-Dimethylmorpholine salt – mp 173-176°, $[\alpha]_D$ + 91° (water) ⟨58-6⟩.

Ethylenediamine salt – mp 230-232° ⟨54-2⟩.

Free acid – (anal, S), mp 210°, UV ⟨39-1⟩.

N-β-Hydroxyethyl-N-methylmorpholine salt – mp 135-142°, $[\alpha]_D$ + 84° (water) ⟨58-6⟩.

N-β-Hydroxyethyl-N-methylpiperidine salt – mp 147-149°, $[\alpha]_D^{24}$ + 85° (water) ⟨58-6⟩.

N-β-Hydroxyethyl-N-methylpyrrolidine salt – mp 148-150°, $[\alpha]_D^{24}$ + 82° ⟨58-6⟩.

N-β-Hydroxyethyltrimethylammonium salt – $C_{23}H_{35}NO_6S$, mp 180-185°, λ_{max} 269, 276 mμ (ε 820 and 750), IR (KBr) ⟨65-19⟩.

Methyl ester – $C_{20}H_{26}O_5S$(?) (anal. C, H, OCH_3, S), sinters 197°, mp 207° ⟨39-1⟩.

N-Methylpyridine salt – mp 152-155°, $[\alpha]_D^{24}$ + 93.3° (water) ⟨58-6⟩.

Piperazine salt – mp 193-195° ⟨53-1⟩.

1-Phenyl-2-aminopropane salt – $C_{27}H_{35}NO_5S$ (anal. C, H, N, S), mp 86-88° ⟨50-3⟩.

Potassium salt – $C_{18}H_{21}KO_5S$ (anal. SO_4), mp 233-242° ⟨38-1⟩ • (anal. sulfated ash), $[\alpha]_D^{20}$ + 114°

(95% ethanol) ⟨53-2⟩ • $C_{18}H_{21}KO_5S \cdot \frac{1}{2}H_2O$ (anal. C, H, K, S, H_2O), mp 219-220°, $[\alpha]_D^{25}$ + 111° (methanol), $\lambda_{max}^{methanol}$ 270 mμ (ε 600) ⟨66-25⟩.

Pyridine salt — $C_{23}H_{37}NO_5S \cdot CHCl_3$ (anal. N, S), mp 170-175°, $[\alpha]_D$ + 84° ⟨57-7⟩ • mp 173-175°, $[\alpha]_D^{23}$ + 84.1° ⟨39-1⟩ • mp 231-233° (recrystallized from methanol-diethyl ether ⟨62-5⟩.

Quinidine salt — $C_{38}H_{46}N_2O_7S \cdot 3H_2O$ (anal. C, H, N), mp 167-170° ⟨39-1⟩, $C_{38}H_{46}N_2O_7S$ (anal. C, H, N, S), $[\alpha]_D^{25}$ + 206° (methanol) ⟨50-2⟩.

Quinine salt — mp 168-170° ⟨39-1⟩.

Sodium salt — $C_{18}H_{21}NaO_5S \cdot H_2O$ (anal. C, H, S), mp 228-230°, $[\alpha]_D^{20}$ + 110°, UV ⟨39-1⟩ • (anal. S, H_2O, sulfated ash), mp sinters 220°, mp 225°, $[\alpha]_D^{20}$ + 104° (water), + 119° (95% ethanol), + 122° (abs. methanol) ⟨53-2⟩ • mp 224-225°, $\lambda_{max}^{methanol}$ 268, 274 mμ (ε 770 and 755), $[\alpha]_D^{23}$ + 115° (water), IR ⟨65-19⟩ • (anal. C, H. Na, S, H_2O), $[\alpha]_D$ + 94° (water), $\lambda_{max}^{50\%\ ethanol}$ 269,275 mμ (ε 780 and 756) ⟨68-4⟩ • mp 226-230° ⟨68-7⟩ • ^{35}S ⟨50-2⟩.

Sodium salt, 17-semicarbazone — mp 258-260° ⟨39-1⟩.

Tetramethylammonium salt — mp 229-230°, $[\alpha]_D^{20}$ + 96.9° (water) ⟨58-6⟩.

Triethylammonium salt — $C_{24}H_{37}NO_5S$ (anal. C, H, N, S), mp 114-116°. $[\alpha]_D^{25}$ + 101° (chloroform), $\lambda_{max}^{methanol}$ 268, 275 mμ (ε 1100 and 1100), NMR ⟨68-2⟩.

Allied Characterization Data

3-SULFATE

CC ⟨62-8⟩⟨65-9⟩⟨66-27⟩⟨67-21⟩⟨67-23⟩ • CCD ⟨61-7⟩ ⟨67-35⟩ • PC ⟨60-11⟩⟨61-10⟩⟨62-8⟩⟨63-14⟩⟨64-1⟩⟨65-22⟩ ⟨67-23⟩ • PE ⟨57-2⟩⟨58-4⟩⟨64-1⟩ • TLC ⟨66-27⟩⟨66-31⟩ ⟨67-23⟩.

3-SULFATE-17-METHOXIME

CC ⟨59-4⟩ • CCD ⟨59-4⟩.

2,3-Dihydroxyestra-1,3,5(10)-trien-17-one
(2-Hydroxyestrone)

3-β-D-GLUCOPYRANOSIDURONIC ACID

CC of urinary metabolite ⟨67-4⟩.

3-β-D-GLUCOPYRANOSIDURONIC ACID-2-METHYL ETHER

CC of urinary metabolite ⟨67-4⟩.

3-SULFATE-2-METHYL ETHER

Potassium salt – $C_{19}H_{23}KO_6S \cdot \frac{1}{2}H_2O$ (anal. C, H, K, S, H_2O), mp 200-250° ⟨66-7⟩ • mp > 260°, $\lambda_{max}^{methanol}$ 222, 282 mμ (ε 8800 and 3200) ⟨66-25⟩.

Sodium salt – mp 170-175° ⟨66-7⟩.

Triethylammonium salt – $C_{25}H_{39}NO_6S$ (anal. C, H, N, S), mp 150-160°, $[\alpha]_D^{25}$ + 101° (chloroform), $\lambda_{max}^{methanol}$ 278 mμ (ε 3390), NMR ⟨68-2⟩.

2,3-DISULFATE

Disodium salt – $C_{18}H_{20}Na_2O_9S_2 \cdot 2H_2O$ (anal. C, H), mp 233-260°, λ_{max} 247 mμ (ε 1740), NMR ⟨68-8⟩.

3,16α-Dihydroxyestra-1,3,5(10)-trien-17-one
(16α-Hydroxyestrone)

16-β-D-GLUCOPYRANOSIDURONIC ACID

Acid – mp 170-175°, impure contains 10-15% 16β-epimer ⟨64-4⟩⟨67-10⟩ • CC ⟨66-12⟩.

Sodium salt – $\lambda_{max}^{methanol}$ 281 mμ, λ_{max}^{NaOH} 295 mμ ⟨61-4⟩.

16-β-D-GLUCOPYRANOSIDURONIC ACID-3-ACETATE

2′,3′,4′-Tri-*O*-acetyl methyl ester – $C_{33}H_{40}O_{13}$ (anal. C,H), mp 240-245°, $[\alpha]_D$ + 75° (chloroform) ⟨65-7⟩⟨67-10⟩.

16-β-D-GLUCOPYRANOSIDURONIC ACID-3-BENZYL ETHER

2′,3′,4′-Tri-*O*-acetyl methyl ester – $C_{38}H_{44}O_{12}$ (anal. C,H), mp 223-226°, $[\alpha]_D$ + 72° (methanol), $\lambda_{max}^{dioxane}$ 278, 287 mμ (ε 1940 and 1800), IR, NMR ⟨67-17⟩ • (anal. C, H), mp 220-223°, $[\alpha]_D$ + 80° (chloroform) ⟨65-7⟩⟨67-10⟩ • (anal. C, H), mp 217.5-220°, $[\alpha]_D^{29.8°}$ + 80° (methanol), $\lambda_{max}^{dioxane}$ 279 and 288 mμ (log ε 3.28 and 3.25) ⟨67-22⟩.

16-β-D-GLUCOPYRANOSIDURONIC ACID-3-METHYL ETHER

2′,3′,4′-Tri-*O*-acetyl methyl ester – $C_{32}H_{40}O_{12}$ (anal. C,H), mp 217-218°, $[\alpha]_D^{21}$ + 71.8° ⟨61-4⟩⟨61-5⟩⟨62-3⟩.

3,16-Di-β-D-GLUCOPYRANOSIDURONIC ACID

Di-2′,3′,4′-tri-*O*-acetyl methyl ester – $C_{44}H_{54}O_{21}$ (anal. C, H), mp 302-306°, $[\alpha]_D^{25}$ + 43.8° (chloroform) ⟨68-1⟩.

Allied Characterization Data

3-GLUCURONIDE

PC ⟨66-13⟩.

16-GLUCURONIDE

CC, PC ⟨67-29⟩.

* * * * *

3-SULFATE-16-β-D-GLUCOPYRANOSIDURONIC ACID

CC ⟨67-29⟩.

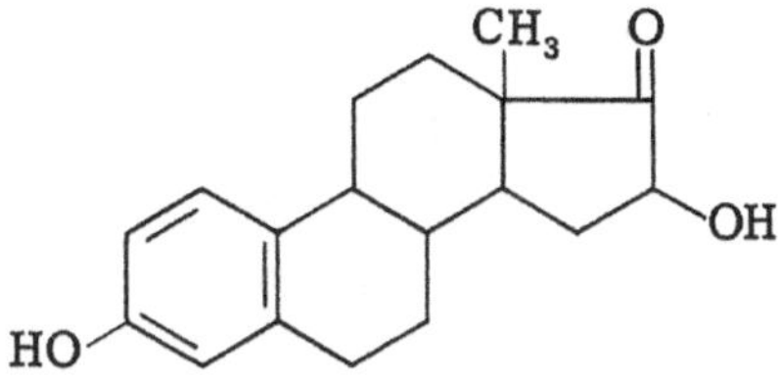

3,16β-Dihydroxyestra-1,3,5(10)-trien-17-one
(16β-Hydroxyestrone)

16-β-D-GLUCOPYRANOSIDURONIC ACID

Acid – $C_{24}H_{30}O_9 \cdot 3H_2O$ (anal. C, H), mp 160-163°, TLC, CC ⟨65-7⟩⟨67-10⟩.

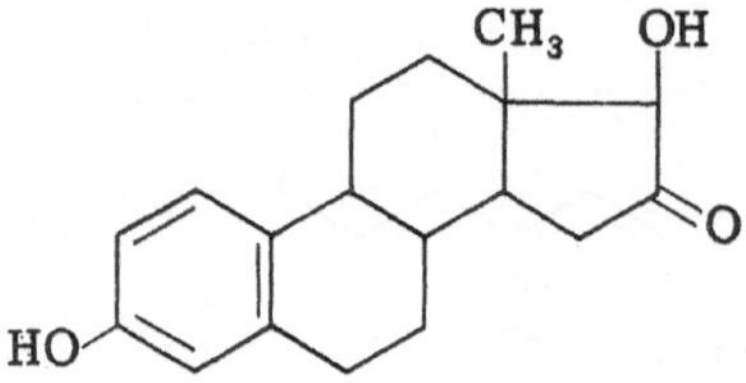

3,17β-Dihydroxyestra-1,3,5(10)-trien-16-one
(16-Keto-17β-estradiol)

17-β-D-GLUCOPYRANOSIDURONIC ACID

Acid – mp 235-245°, CC ⟨67-10⟩.

17-β-D-GLUCOPYRANOSIDURONIC ACID-3-BENZYL ETHER

2′,3′,4′-Tri-*O*-acetyl methyl ester – $C_{38}H_{44}O_{12}$, mp 152-154° purity (?) ⟨67-10⟩.

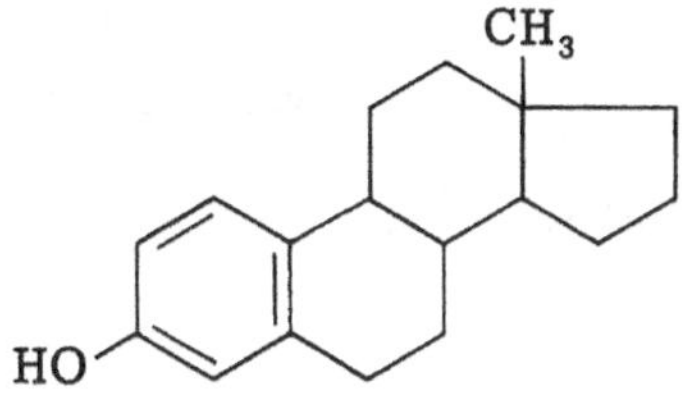

Estra-1,3,5(10)-trien-3-ol

3-SULFATE

Triethylammonium salt – $C_{24}H_{39}NO_4S$ (anal. C, H, N, S), mp 135-136°, $[\alpha]_D^{25}$ + 47° (chloroform), $\lambda_{max}^{methanol}$ 270 and 277 mμ (ε 987 and 940), NMR ⟨68-2⟩.

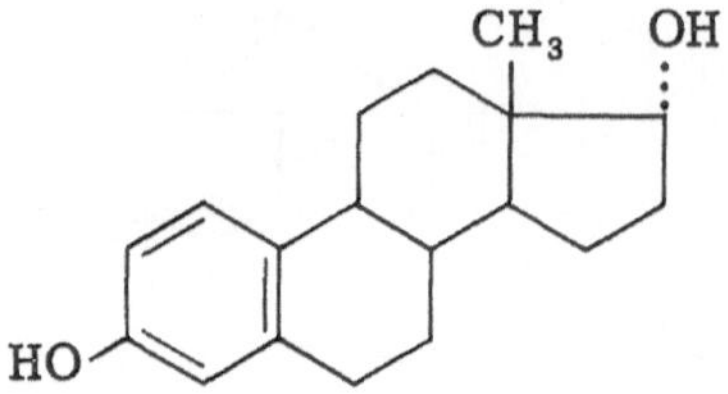

Estra-1,3,5(10)-triene-3,17α-diol
(17α-Estradiol)

3-β-D-GLUCOPYRANOSIDURONIC ACID

Acid – mp 155.5-156.5°, $[\alpha]_D$ - 23.2° (ethanol), CCD, TLC ⟨67-3⟩ • PC, TLC ⟨66-33⟩.

3-β-D-GLUCOPYRANOSIDURONIC ACID-17-2′-ACETAMIDO-2′-DEOXY-β-D-GLUCOPYRANOSIDE

IR, CCD, 16-^{14}C ⟨65-15⟩ • CCD, 6,7-^{3}H ⟨65-12⟩ • CCD, TLC ⟨67-3⟩.

17-2′-AMINO-2′-DEOXY-β-D-GLUCOPYRANOSIDE

2′-Acetyl-$C_{26}H_{37}NO_7 \cdot H_2O$ (anal. C, H, N, O), mp 227-229°, $\lambda_{max}^{methanol}$ 280, 286 (shoulder) mμ, IR, TLC ⟨64-13⟩ • mp 225-229°, $[\alpha]_D$ - 16.2° (ethanol), CCD, TLC ⟨67-3⟩ • 6,7-^{3}H, CCD, TLC ⟨65-12⟩.

2′,3′,4′,6′-Tetraacetyl – IR, TLC ⟨64-13⟩.

3-SULFATE

Sodium salt – $C_{18}H_{23}NaO_5S \cdot H_2O$ ⟨50-2⟩.

17-SULFATE

CC ⟨65-5⟩.

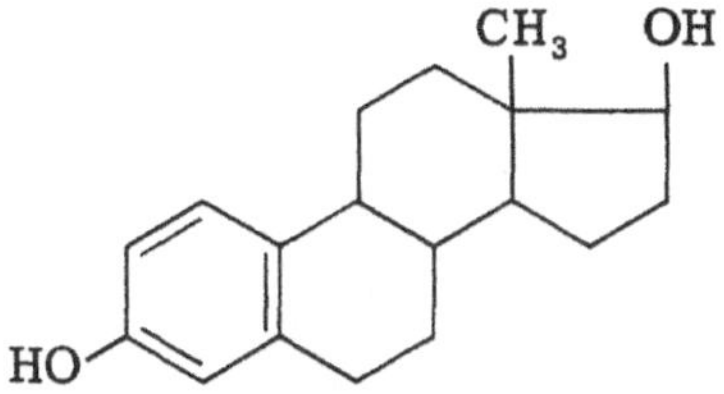

Estra-1,3,5(10)-triene-3,17β-diol
(17β-Estradiol, Estradiol)

17-α-D-GLUCOFURANOSIDURONIC ACID

γ-Lactone – $C_{24}H_{30}O_7 \cdot CH_3CO_2C_2H_5$ (anal. C, H), mp 270-273°, $[\alpha]_D^{25}$ + 85° (dimethylformamide), IR ⟨63-16⟩ • $C_{24}H_{30}O_7$, IR ⟨66-17⟩.

3-β-D-GLUCOPYRANOSIDURONIC ACID

Sodium salt – $C_{24}H_{31}NaO_8 \cdot 3H_2O$ (anal. C, H), mp 255-258° ⟨67-10⟩.

2′,3′,4′-Tri-*O*-acetyl methyl ester – $C_{31}H_{40}O_{11}$ (anal. C, H), mp 207-209°, $[\alpha]_D^{25}$ + 14° (chloroform) ⟨67-10⟩.

17-β-D-GLUCOPYRANOSIDURONIC ACID

Acid – $C_{24}H_{32}O_8$ (anal. C, H), mp 198-202°, $[\alpha]_D^{24}$ −4° (0.5 M sodium hydroxide) ⟨59-7⟩ • mp 191-194° ⟨38-3⟩ • (anal. C, H), mp 191-194°, IR, TLC ⟨66-17⟩ • $C_{24}H_{32}O_8 \cdot 1\frac{1}{2}H_2O$, mp 191-194° ⟨39-3⟩ • $C_{24}H_{32}O_8 \cdot CH_3OH$ (anal. C, H), mp 190-193°, TLC, PC ⟨64-4⟩ ⟨65-7⟩ ⟨67-10⟩.

Amide – $C_{24}H_{33}NO_7$ (anal. C, H), mp 214-217°, $[\alpha]_D^{25}$ − 11° (dimethylformamide), IR ⟨63-16⟩.

Barium salt – no physical contants ⟨39-3⟩.

γ-Lactone – $C_{24}H_{30}O_7 \cdot CH_3OH$ (anal. C, H), mp 208-211°, $[\alpha]_D^{25}$ 0° (dimethylformamide), IR ⟨63-16⟩.

Methyl ester – $C_{25}H_{34}O_8 \cdot CH_3OH$ (anal. C, H), mp 132-135°, IR ⟨65-7⟩.

Potassium salt – mp 286-288° ⟨66-6⟩.

Sodium salt – $C_{24}H_{31}NaO_8 \cdot 2H_2O$ (anal. C, H), mp 290-293°, CC, PC ⟨64-4⟩⟨65-7⟩⟨67-10⟩ • 6,7-^{3}H, CC ⟨66-17⟩.

2′,3′,4′-Tri-*O*-acetyl methyl ester – $C_{31}H_{40}O_{11}$ (anal. C, H), mp 120-123°, NMR ⟨66-6⟩ • $C_{31}H_{40}O_{11} \cdot H_2O$ (anal. C,H), mp 122-124°, $[\alpha]_D^{25}$ – 7° ⟨59-7⟩ • IR ⟨59-8⟩ • mp 121-123° ⟨64-4⟩⟨65-7⟩⟨67-10⟩.

17-β-D-GLUCOPYRANOSIDURONIC ACID-3-ACETATE

Acid – $C_{31}H_{40}O_{11}$ (anal. C, H), mp 195-195.5°, $[\alpha]_D^{21.5}$ + 2.5° (ethanol) IR ⟨61-4⟩.

17-β-D-GLUCOPYRANOSIDURONIC ACID-3-BENZOATE

2′,3′,4′-Tri-*O*-acetyl methyl ester – $C_{38}H_{44}O_{12}$, mp 188-191.5°, $[\alpha]_D^{20}$ + 9.2° (chloroform) ⟨38-3⟩⟨39-3⟩.

17-β-D-GLUCOPYRANOSIDURONIC ACID-3-BENZYL ETHER

2′,3′,4′-Tri-*O*-acetyl methyl ester – $C_{38}H_{46}O_{11}$ (anal. C, H), mp 199.5-200.5°, $[\alpha]_D^{25}$ + 3.7° (chloroform), NMR, $\lambda_{max}^{methanol}$ 278, 287 mμ (ε 5430 and 4570) ⟨66-6⟩ • (anal. C, H), mp 199-202°, $[\alpha]_D$ + 12° (chloroform), IR ⟨64-4⟩⟨65-7⟩⟨67-10⟩.

17-β-D-GLUCOPYRANOSIDURONIC ACID-3-SULFATE

Dipotassium salt – $C_{24}H_{30}K_2O_{11}S$ (anal. C, H, K, S), mp 246-257°, $[\alpha]_D$ – 17° (water), λ_{max}^{water} 268, 274 mμ (ε 790 and 750) NMR, PE ⟨66-6⟩.

3,17-Di-β-D-GLUCOPYRANOSIDURONIC ACID

Di-2′,3′,4′-Tri-*O*-acetyl methyl ester — $C_{44}H_{56}O_{20}$ (anal. C, H), mp 215-216.5°, $[\alpha]_D^{25}$ − 5° (methanol) ⟨59-7⟩ • IR ⟨59-8⟩.

Allied Characterization Data

3-GLUCURONIDE

CCD ⟨61-7⟩⟨62-4⟩, CCD, 6,7-^{3}H, 16-^{14}C ⟨67-35⟩, PC ⟨66-13⟩ ⟨66-33⟩ • TLC ⟨66-2⟩⟨66-33⟩.

17-GLUCURONIDE

CC ⟨61-4⟩⟨63-2⟩⟨66-8⟩⟨66-17⟩⟨67-21⟩⟨67-29⟩ • CCD ⟨61-4⟩ ⟨61-7⟩⟨62-4⟩, PC ⟨61-4⟩⟨66-12⟩⟨66-13⟩, TLC ⟨65-8⟩⟨66-2⟩ ⟨66-8⟩⟨66-17⟩⟨67-29⟩.

3,17-DIGLUCURONIDE

CC ⟨63-2⟩, CCD ⟨61-7⟩⟨62-4⟩, TLC ⟨66-2⟩.

* * * * *

3-SULFATE

Piperazine salt — mp 188° ⟨53-1⟩.

Potassium salt — $C_{18}H_{23}KO_5S$ (anal. K) ⟨68-4⟩.

Sodium salt — $C_{18}H_{23}NaO_5S \cdot H_2O$ (anal. C, H, Na, S, H_2O), $[\alpha]_D$ + 46° (water), $\lambda_{max}^{50\%\ ethanol}$ 269 and 275 mμ (ε 783 and 768) ⟨68-4⟩ • $C_{18}H_{23}NaO_5S \cdot H_2O$ ⟨50-2⟩ • $C_{18}H_{23}NaO_5S \cdot H_2O \cdot CH_3OH$ (anal. C, H, Na, S), mp 202-212°, $\lambda_{max}^{methanol}$ 270 and 276 mμ (ε 270 and 770), $[\alpha]_D$ + 41° (water), IR (KBr) ⟨65-13⟩ • PC ⟨61-10⟩.

3-SULFATE-17-ACETATE

Sodium salt – $C_{20}H_{25}NaO_6S{\cdot}H_2O$ (anal. acetyl), $[\alpha]_D$ + 14° (33% ethanol), $\lambda_{max}^{50\%\ ethanol}$ 269 and 275 mμ (ε 771 and 696) ⟨68-4⟩.

Triethylammonium salt – $C_{26}H_{41}NO_6S$ (anal. C, H, N, S), mp 142-145°, $[\alpha]_D^{25}$ + 31° (chloroform), $\lambda_{max}^{methanol}$ 269 and 275 mμ, NMR ⟨68-3⟩.

17-SULFATE

Sodium salt – $C_{18}H_{23}NaO_5S{\cdot}H_2O$ (anal. C, H, Na), mp indefinite 160°-180°, $\lambda_{max}^{alcohol}$ 280 mμ (ε 2000), λ_{max}^{water} 276-278 mμ (ε 1770), $[\alpha]_D^{25}$ + 16° (water), $[\alpha]_D^{25}$ + 42.5° (alcohol) ⟨50-5⟩ • $C_{18}H_{23}NaO_5S{\cdot}2H_2O$ (anal. C, H, S, H_2O), $[\alpha]_D$ + 35° (33% ethanol), $\lambda_{max}^{50\%\ ethanol}$ 279 mμ (ε 1978) ⟨68-4⟩ • $C_{18}H_{23}NaO_5S{\cdot}4H_2O$ (anal. C, H, Na, S), mp 188-190°, $\lambda_{max}^{methanol}$ 281 mμ (ε 2220), $[\alpha]_D$ + 20° (water), IR, CCD, PC ⟨65-13⟩.

Triethylammonium salt – $C_{24}H_{39}NO_5S$ (anal. C, H, N, S), mp 200-201°, $[\alpha]_D^{25}$ + 4° (chloroform), $\lambda_{max}^{methanol}$ 281 mμ (ε 2260), ⟨68-3⟩.

17-SULFATE-3-ACETATE

Triethylammonium salt – $C_{26}H_{41}NO_6S$ (anal. C, H, N, S), mp 171-172°, $[\alpha]_D^{25}$ + 31° (chloroform), $\lambda_{max}^{methanol}$ 268 and 275 mμ ⟨68-3⟩.

17-SULFATE-3-BENZOATE

Pyridine salt – $C_{30}H_{33}NO_6S{\cdot}H_2O$ (anal. C, H, N, O, S), mp 162-167°, $\lambda_{max}^{methanol}$ 230 mμ (ε 20,490), $[\alpha]_D$ + 50° (chloroform), IR (chloroform) ⟨65-13⟩.

Sodium salt – $C_{25}H_{27}NaO_6S{\cdot}\frac{1}{2}H_2O$ (anal. C, H, Na), mp 155-195°, $[\alpha]_D^{25}$ + 34.8° (alcohol) ⟨50-5⟩.

Triethylammonium salt – $C_{31}H_{43}NO_6S$ (anal. C, H, N, S), mp 201-202°, $[\alpha]_D^{25}$ + 25° (chloroform), $\lambda_{max}^{methanol}$ 229, 267 and 275 mμ (ε 20,900, 5000 and 4180) ⟨68-3⟩.

17-SULFATE-3-METHYL ETHER

Triethylammonium salt – $C_{25}H_{41}NO_5S$ (anal. C, H, N, S), mp 176-177°, $[\alpha]_D^{25}$ + 33° (chloroform), $\lambda_{max}^{methanol}$ 279 and 287 mμ (ε 2100 and 1870), NMR ⟨68-2⟩.

3,17-DISULFATE

Diammonium salt – $C_{18}H_{30}N_2O_8S_2 \cdot H_2O$ (anal. C, H, N, S, H_2O), mp 185-190°, $[\alpha]_D^{25}$ + 28.4° (methanol), $\lambda_{max}^{methanol}$ 269 and 278 mμ (ε 1030 and 900) ⟨68-3⟩.

Dipotassium salt – $C_{18}H_{22}K_2O_8S_2$ (anal. C, H, ash) ⟨60-13⟩ • (anal. K, S) ⟨68-4⟩.

Diquinidine salt – $C_{58}H_{73}N_4O_{12}S_2 \cdot 2H_2O$, $[\alpha]_D^{20}$ + 183° (methanol) ⟨50-2⟩.

Disodium salt – $C_{18}H_{22}Na_2O_8S_2 \cdot 2CH_3OH$ (anal. C, H, Na, S), mp 186-191° (dec), $\lambda_{max}^{methanol}$ 270, 276 mμ (ε 1000 and 975), $[\alpha]_D$ + 33° (water), IR (KBr), 6,7-^{3}H, CCD, PC ⟨65-13⟩ • $C_{18}H_{22}Na_2O_8S_2 \cdot 2H_2O$ (anal. C, H, Na, S, H_2O), $[\alpha]_D$ + 33° (water), $\lambda_{max}^{50\%\ ethanol}$ 269, 275 mμ (ε 825 and 762) ⟨68-4⟩.

Allied Characterization Data

3-SULFATE

CC ⟨63-2⟩ • CCD ⟨61-7⟩⟨61-10⟩⟨62-4⟩⟨67-7⟩⟨67-35⟩ • PE ⟨61-2⟩ • TLC ⟨62-4⟩ • PC ⟨61-2⟩⟨63-12⟩⟨65-22⟩ ⟨67-7⟩.

17-SULFATE

CC ⟨63-2⟩⟨65-9⟩ • CCD ⟨61-7⟩⟨62-4⟩ • PC ⟨63-14⟩ ⟨65-5⟩⟨65-22⟩⟨67-7⟩ • TLC ⟨62-4⟩.

3,17-DISULFATE

CC ⟨63-2⟩ ⟨67-21⟩ • CCD ⟨61-7⟩ ⟨62-4⟩ • TLC ⟨62-4⟩ • PC ⟨63-14⟩.

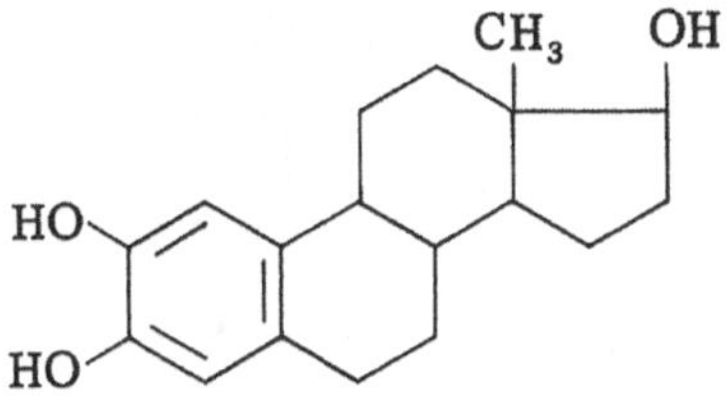

Estra-1,3,5(10)-triene-2,3,17β-triol
(2-Hydroxy-17β-estradiol, 2-Hydroxyestradiol)

2-SULFATE

Sodium salt – $C_{18}H_{23}NaO_6S \cdot 2\frac{1}{2}H_2O$ (anal. C, H), mp 206-209°, $\lambda_{max}^{ethanol}$ 281 mμ (ε 2700), NMR, TLC ⟨68-8⟩.

3-SULFATE

Sodium salt – $C_{18}H_{23}NaO_6S \cdot 1\frac{1}{2}H_2O$ (anal. C, H), mp 192-193°, $\lambda_{max}^{ethanol}$ 282 mμ (ε 2730), NMR, TLC ⟨68-8⟩.

2-SULFATE-3-(2′-BENZOYL-4′-NITRO)PHENYL ETHER-17-ACETATE

Trimethylammonium salt – mp 191-194°, $\lambda_{max}^{ethanol}$ 257 mμ, TLC ⟨68-8⟩.

3-SULFATE-2-(2′-BENZOYL-4′-NITRO)PHENYL ETHER-17-ACETATE

Trimethylammonium salt – mp 190-194°, TLC ⟨68-8⟩.

Allied Characterization Data

3-SULFATE-2-METHYL ETHER

CCD ⟨61-7⟩.

Estra-1,3,5(10)-triene-3,15α,17β-triol
(15α-Hydroxy-17β-estradiol, 15α-Hydroxyestradiol)

3-SULFATE

⟨65-25⟩.

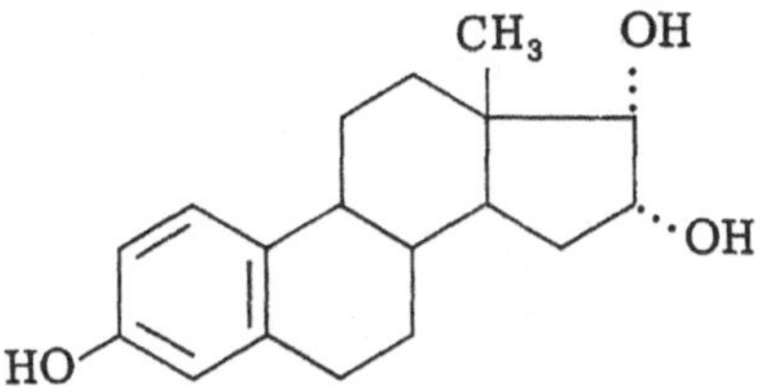

Estra-1,3,5(10)-triene-3,16α,17α-triol
(17-Epiestriol)

16-β-D-GLUCOPYRANOSIDURONIC ACID

Sodium salt – $C_{24}H_{31}NaO_9 \cdot 5H_2O$ (anal. C, H), mp 240-245°, CC, TLC ⟨65-7⟩ ⟨67-10⟩.

2′,3′,4′-Tri-*O*-acetyl methyl ester – TLC ⟨65-7⟩.

16-β-D-GLUCOPYRANOSIDURONIC ACID-3-BENZYL ETHER

2′,3′,4′-Tri-*O*-acetyl methyl ester – $C_{38}H_{46}O_{12}$ (anal. C, H), mp 200-204.5°, $[\alpha]_D^{16.4}$ + 23.0 (chloroform), $\lambda_{max}^{dioxane}$ 278 and 286 mμ (log ε 3.30 and 3.27), TLC ⟨67-22⟩.

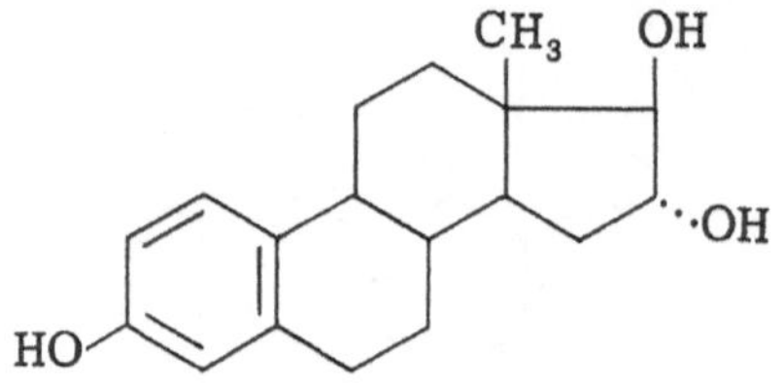

Estra-1,3,5(10)-triene-3,16α,17β-triol
(Estriol)

3-β-D-GLUCOPYRANOSIDURONIC ACID

Acid – $C_{24}H_{32}O_9$ (anal. C, H), mp 213-220° (dec), $[\alpha]_D^{26.6}$ −18.1° (ethanol), $\lambda_{max}^{ethanol}$ 274 and 282 mμ (log ε 3.22 and 3.15) ⟨67-22⟩.

Sodium salt – $C_{24}H_{31}NaO_9 \cdot 3H_2O$ (anal. C, H), mp 269-271°, mp 272.5-281° (aqueous ethanol) ⟨67-10⟩.

3-β-D-GLUCOPYRANOSIDURONIC ACID-16,17-DIACETATE

2′,3′,4′-Tri-*O*-acetyl methyl ester – $C_{35}H_{44}O_{14}$ (anal. C,H), mp 192-194°, $[\alpha]_D^{25}$ −31° (chloroform) ⟨67-10⟩ • (anal. C, H), mp 197-198°, $[\alpha]_D^{25.8°}$ − 0.2° (chloroform), $\lambda_{max}^{dioxane}$ 274 and 282 mμ (log ε 3.11 and 3.04), TLC ⟨67-22⟩.

16-β-D-GLUCOPYRANOSIDURONIC ACID

Acid – $C_{24}H_{32}O_9$ (anal. C, H), mp 221-222°, $\lambda_{max}^{ethanol}$ 280 mμ (ε 2100), $\lambda_{max}^{0.5N\ sodium\ hydroxide\text{-}ethanol}$ 300 mμ ⟨61-11⟩ • (anal. C, H), mp 223-224°, 226° (immersed at 200°), $[\alpha]_D^{24}$ + 1.5° (95% ethanol) ⟨63-8⟩ • $C_{24}H_{32}O_9 \cdot 3H_2O$ (anal. C, H), mp 217-225° ⟨64-4⟩⟨67-10⟩ • 16-^{14}C ⟨66-19⟩⟨66-20⟩ • $C_{24}H_{34}O_9 \cdot H_2O$ (anal. C, H), mp 224-225°, $\lambda_{max}^{dioxane}$ 281 mμ (log ε 3.29), $[\alpha]_D^{26.5}$ ± 0° (ethanol) ⟨67-22⟩.

Sodium salt – $C_{24}H_{31}NaO_9$ (anal. C, H), mp 246-249° ⟨65-7⟩ ⟨67-10⟩ • mp 243°, $\lambda_{max}^{ethanol}$ 280 mμ ⟨61-11⟩ • mp 247-248°, λ_{max} 280 mμ (ε 1900) ⟨63-8⟩ • $C_{23}H_{31}NaO_9 \cdot 2H_2O$ (anal. C, H), mp 246-252° ⟨67-22⟩.

Tri-*O*-acetyl methyl ester — $C_{31}H_{40}O_{12}$ (anal. C, H), mp 228-229° (dec), $[\alpha]_D^{17.6°}$ + 17.6° (chloroform), $\lambda_{max}^{dioxane}$ 281 and 289 mμ (log ϵ 3.27 and 3.23) ⟨67-22⟩ • TLC ⟨65-7⟩.

16-β-D-GLUCOPYRANOSIDURONIC ACID-3-BENZYL ETHER

2′,3′,4′-Tri-*O*-acetyl methyl ester — $C_{38}H_{46}O_{12}$ (anal. C,H), mp 245-246°, $[\alpha]_D$ + 5.8° (methanol), $\lambda_{max}^{methanol}$ 278 and 287 mμ (ϵ 2415 and 2250) ⟨67-17⟩ • (anal. C, H), mp 244-247° (dec), $[\alpha]_D^{30}$ + 15° (chloroform), $\lambda_{max}^{dioxane}$ 278 and 287 mμ (log ϵ 3.27 and 3.24), TLC ⟨67-22⟩.

16-β-D-GLUCOPYRANOSIDURONIC ACID-3-METHYL ETHER

Methyl ester — $\lambda_{max}^{ethanol}$ 278 and 286 mμ ⟨61-11⟩⟨62-6⟩.

2′,3′,4′-Tri-*O*-acetyl methyl ester — $C_{34}H_{42}O_{16}$ (anal. C,H), mp 214-216.5°, $[\alpha]_D^{21}$ + 12.7° (ethanol), IR ⟨61-4⟩⟨62-3⟩.

16-β-D-GLUCOPYRANOSIDURONIC ACID-3-SULFATE

Dipotassium salt — $C_{24}H_{30}K_2O_{12}S \cdot 2H_2O$ (anal. C, H, K, S, H_2O), mp > 250°, $[\alpha]_D^{25}$ − 5.3° (water), $\lambda_{max}^{methanol}$ 268 and 270 mμ (ϵ 1000 and 1100) ⟨67-17⟩.

16-β-D-GLUCOPYRANOSIDURONIC ACID-17-ACETATE

2′,3′,4′-Tri-*O*-acetyl methyl ester — $C_{33}H_{42}O_{13}$ (anal. C,H), mp 213-215°, $[\alpha]_D$ − 30.4° (methanol), $\lambda_{max}^{methanol}$ 280 and 287 mμ (ϵ 2070 and 1800) ⟨67-17⟩.

16-β-D-GLUCOPYRANOSIDURONIC ACID-3-BENZYL ETHER-17-ACETATE

2′,3′,4′-Tri-*O*-acetyl methyl ester — $C_{40}H_{48}O_{13}$ (anal. C, H), mp 180°, $[\alpha]_D$ − 23° (methanol), $\lambda_{max}^{methanol}$ 278 and 287 mμ (ϵ 2500 and 2100) ⟨67-17⟩.

16-β-D-GLUCOPYRANOSIDURONIC ACID-3-METHYL ETHER-17-ACETATE

2′,3′,4′-Tri-*O*-acetyl methyl ester — $C_{34}H_{44}O_{13}$ (anal. C,H), mp 165.5-167°, 180-182°, $[\alpha]_D^{21}$ − 22.4° (ethanol) ⟨61-5⟩ ⟨62-3⟩ • sinters 166-167°, mp 185.5-186°, $\lambda_{max}^{ethanol}$ 278 and 287 mμ (ε 1950 and 1700), IR ⟨62-6⟩ ⟨63-8⟩ • (anal. C, H), mp 165-166° melts, resolidifies 179-182° remelts, $[\alpha]_D^{25}$ − 17.1° (ethanol), $\lambda_{max}^{methanol}$ 278 and 287 mμ (ε 1800 and 1700) ⟨67-17⟩ • CC, PC ⟨61-4⟩.

16-β-D-GLUCOPYRANOSIDURONIC ACID-3,17-DIMETHYL ETHER

2′,3′,4′-Tri-*O*-methyl ester — $C_{30}H_{44}O_9$ (anal. C, H, OCH_3), mp 143-144°, $[\alpha]_D^{24}$ − 2.1° (chloroform), $\lambda_{max}^{methanol}$ 278 and 287 mμ (ε 1800 and 1620), IR ⟨62-6⟩ ⟨63-8⟩ • mp 140-141°, $\lambda_{max}^{ethanol}$ 279 and 287 mμ (ε 1800 and 1620), IR ⟨61-11⟩.

16-β-D-GLUCOPYRANOSIDURONIC ACID-3-SULFATE-17-ACETATE

2′,3′,4′-Tri-*O*-acetyl methyl ester-triethylammonium salt — $C_{39}H_{57}NO_{16}S$ (anal. C, H, N, S), mp 212-213°, $[\alpha]_D^{25}$ −27.7° (methanol), $\lambda_{max}^{methanol}$ 267 and 277 mμ (ε 868 and 828) ⟨67-17⟩.

17-β-D-GLUCOPYRANOSIDURONIC ACID

Acid — $C_{24}H_{32}O_9$, mp 253-256° ⟨64-4⟩ ⟨67-10⟩ • mp 262-263° ⟨65-7⟩ • $C_{24}H_{32}O_9 \cdot H_2O$ (anal. C, H), mp 235-240°, (dec), $[\alpha]_D^{16.5°}$ − 116° (methanol), $\lambda_{max}^{ethanol}$ 281 mμ (log ε 3.29) ⟨67-22⟩.

Sodium salt — $C_{24}H_{31}NaO_9 \cdot 3H_2O$ (anal. C, H), mp > 310° ⟨67-10⟩ • PC, TLC ⟨65-7⟩.

2′,3′,4′-Tri-*O*-acetyl methyl ester — $C_{31}H_{40}O_{12}\cdot H_2O$ (anal. C, H), mp 207-212°, $[\alpha]_D^{25.4°}$ − 64° (chloroform), $\lambda_{max}^{dioxane}$ 281 and 288 mμ (log ε 3.35 and 3.30) ⟨67-22⟩.

17-β-D-GLUCOPYRANOSIDURONIC ACID-3-BENZYL ETHER

2′,3′,4′-Tri-*O*-acetyl methyl ester — $C_{38}H_{46}O_{12}$ (anal. C,H), mp 183-184.5°, $[\alpha]_D^{21}$ + 0.86° (chloroform), $\lambda_{max}^{dioxane}$ 279 and 287 mμ (log ε 3.30 and 3.26), TLC ⟨67-22⟩.

17-β-D-GLUCOPYRANOSIDURONIC ACID-3-BENZYL ETHER-16-ACETATE

2′,3′,4′-Tri-*O*-acetyl methyl ester — $C_{40}H_{48}O_{13}$ (anal. C,H), mp 202-204.5°, $[\alpha]_D^{21.5}$ − 21° (ethanol) ⟨61-4⟩ ⟨67-10⟩.

17-β-D-GLUCOPYRANOSIDURONIC ACID-3-BENZYL ETHER-16-*t*-BUTYL ETHER

2′,3′,4′-Tri-*O*-acetyl methyl ester — $C_{42}H_{54}O_{12}$ (anal. C,H), mp 153-156°, $[\alpha]_D^{27.8}$ − 289° (chloroform), $\lambda_{max}^{dioxane}$ 278 and 286 mμ (log ε 3.31 and 3.28), TLC ⟨67-22⟩.

17-β-D-GLUCOPYRANOSIDURONIC ACID-3-METHYL ETHER-16-ACETATE

2′,3′,4′-Tri-*O*-acetyl methyl ester — $C_{34}H_{44}O_{13}$ (anal. C,H), mp 169-170°, $[\alpha]_D^{21}$ − 21.2° (ethanol) ⟨61-5⟩ ⟨62-3⟩ • CC, PC ⟨61-4⟩.

MONO-β-D-GLUCOPYRANOSIDURONIC ACID (Unspecified Location)

Acid — $C_{24}H_{32}O_9$ (anal. C, H), mp 199-201° ⟨50-1⟩ • $C_{24}H_{32}O_9$, mp 224-226°, $[\alpha]_D^{17}$ − 7.5° (ethanol) ⟨50-4⟩ • (anal. C, H), mp 193-197° ⟨35-1⟩ ⟨36-1⟩.

Barium salt – $C_{24}H_{31}O_9Ba\frac{1}{2}$ (anal. Ba) ⟨35-1⟩⟨36-1⟩.

Sodium salt – $C_{24}H_{31}NaO_9 \cdot 1\frac{1}{2}H_2O$ (anal.), mp 256-257° • $C_{24}H_{31}NaO_9 \cdot \frac{1}{2}CH_3OH$ (anal.), mp 256-257° • $C_{24}H_{31}NaO_9 \cdot \frac{1}{2}CH_3OH$ (anal.), mp 305-306°, $[\alpha]_D^{28}$ – 28.2° (water) – 21.0° (water) ⟨36-2⟩ • $C_{24}H_{31}NaO_9$ (anal. C, H, Na), mp 252-256°, mp 282-283° ⟨50-1⟩ • $C_{24}H_{31}NaO_9$ (anal. C, H, Na), mp 245-248.5° ⟨50-4⟩.

Allied Characterization Data

3-GLUCURONIDE

CC ⟨63-2⟩⟨66-20⟩⟨66-23⟩⟨67-13⟩⟨67-29⟩ • CCD ⟨66-19⟩⟨66-20⟩⟨66-44⟩ • PC ⟨66-11⟩⟨66-12⟩⟨67-29⟩, TLC ⟨67-25⟩⟨67-29⟩.

16-GLUCURONIDE

CC ⟨65-7⟩⟨65-9⟩⟨67-13⟩⟨67-25⟩⟨67-29⟩ • CCD ⟨61-15⟩⟨64-25⟩⟨65-16⟩⟨66-19⟩⟨66-44⟩⟨67-11⟩ • PC ⟨64-25⟩⟨65-16⟩⟨65-17⟩⟨66-11⟩⟨66-12⟩⟨66-13⟩⟨67-11⟩⟨67-25⟩⟨67-29⟩ • PE ⟨61-15⟩⟨64-11⟩ • TLC ⟨65-7⟩⟨67-25⟩⟨67-29⟩.

16-GLUCURONIDE-3-SULFATE

CC ⟨65-16⟩⟨65-17⟩⟨67-29⟩ • CCD ⟨65-16⟩⟨65-17⟩⟨66-20⟩⟨67-11⟩ • PC ⟨65-16⟩⟨65-17⟩ • PE ⟨65-16⟩⟨65-17⟩.

17-GLUCURONIDE

CC ⟨67-29⟩ • PC ⟨66-11⟩⟨66-12⟩⟨66-13⟩⟨67-29⟩.

GLUCURONIDE (Unspecified Location)

⟨65-8⟩

DIGLUCURONIDE (Unspecified Location)

CC ⟨61-8⟩ • PC ⟨61-8⟩.

* * * * *

3-SULFATE

Acid – 15β-^{3}H ⟨66-20⟩.

Potassium salt – $C_{18}H_{23}KO_6S$ (anal. K, S) ⟨68-4⟩.

Sodium salt – $C_{18}H_{23}NaO_6S$, mp 218-222°, IR ⟨65-16⟩ • $C_{18}H_{23}NaO_6S{\cdot}H_2O$ (anal. C, H, Na, S, H_2O), $[\alpha]_D$ + 36° (water), $\lambda_{max}^{50\%\ ethanol}$ 269 and 275 mμ (ε 790 and 737) ⟨68-4⟩.

3-SULFATE-16,17-DIACETATE

Sodium salt – $C_{22}H_{27}NaO_8S{\cdot}H_2O$ (anal. C, H, S, acetyl), $[\alpha]_D$ – 9° (water), $\lambda_{max}^{50\%\ ethanol}$ 269 and 275 mμ (ε 711 and 664), IR, PC ⟨68-4⟩ • CCD ⟨61-7⟩.

17-SULFATE

Sodium salt – $C_{18}H_{23}NaO_6S{\cdot}3H_2O$ (anal. C, H, S), $[\alpha]_D$ + 6° (water), $\lambda_{max}^{50\%\ ethanol}$ 279 mμ (ε 2005), IR, PC ⟨68-4⟩.

17-SULFATE-3,16-DIACETATE

Sodium salt – $C_{22}H_{27}NaO_8S{\cdot}H_2O$ (anal. C, H, S, acetyl), $\lambda_{max}^{50\%\ ethanol}$ 269 and 275 mμ (ε 699 and 645), ⟨68-4⟩.

3,17-DISULFATE

Disodium salt — $C_{18}H_{22}Na_2O_9S_2 \cdot 2H_2O$ (anal. S), $[\alpha]_D$ + 11° (water), $\lambda_{max}^{50\%\ ethanol}$ 269 and 275 mμ (ε 775 and 717) ⟨68-4⟩.

16,17-DISULFATE

Disodium salt — $C_{18}H_{22}Na_2O_9S_2 \cdot 2H_2O$ (anal. C, H, S, Na, H_2O), $[\alpha]_D$ − 11° (water), $\lambda_{max}^{50\%\ ethanol}$ 279 mμ (ε 1965) ⟨68-4⟩ • CCD ⟨61-7⟩.

3,16,17-TRISULFATE

Triammonium salt — $C_{18}H_{33}N_3O_{12}S_3$ (anal. C, H, N, S), mp 322-325°, $[\alpha]_D^{25}$ − 11° (dimethylsulfoxide), $\lambda_{max}^{methanol}$ 270 and 276 mμ (ε 1420 and 1000) ⟨68-3⟩.

Tripotassium salt — $C_{18}H_{21}K_3O_{12}S_3$ (anal. C, H, K, S), $[\alpha]_D$ − 7° (water), $\lambda_{max}^{50\%\ ethanol}$ 269 and 275 mμ (ε 813 and 776) ⟨68-4⟩.

Trisodium salt — $C_{18}H_{21}Na_3O_{12}S_3$ ⟨50-2⟩.

Allied Characterization Data

3-SULFATE

CC ⟨63-2⟩ ⟨65-9⟩ • CCD ⟨61-7⟩ ⟨61-10⟩ ⟨61-15⟩ ⟨64-26⟩ ⟨66-20⟩ ⟨66-44⟩ • PC ⟨61-10⟩ ⟨61-15⟩ ⟨65-16⟩ ⟨67-7⟩ ⟨67-11⟩ • PE ⟨61-15⟩.

3-SULFATE-16,17-DIACETATE

PC ⟨61-10⟩ • CCD ⟨61-15⟩ ⟨66-44⟩.

SULFATE (Unspecified Location)

CC ⟨61-2⟩ • PE ⟨61-2⟩⟨61-10⟩.

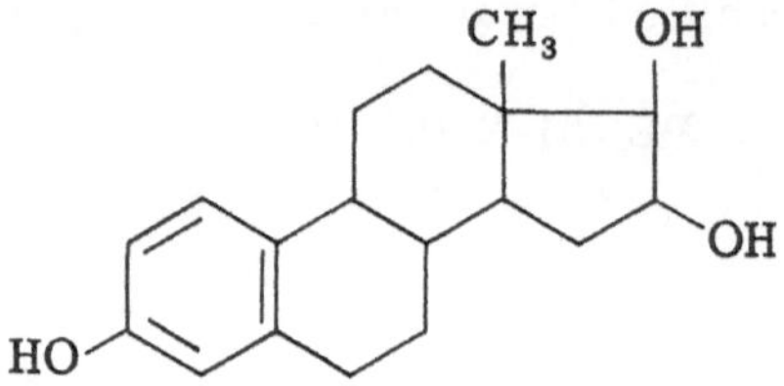

Estra-1,3,5(10)-triene-3,16β,17β-triol
(16-Epiestriol)

16-β-D-GLUCOPYRANOSIDURONIC ACID

Acid – No physical constants ⟨64-4⟩.

16-β-D-GLUCOPYRANOSIDURONIC ACID-3-BENZYL ETHER-17-ACETATE

Acid – mp 165-170°, 172-183°, CC, TLC ⟨65-7⟩.

17-β-D-GLUCOPYRANOSIDURONIC ACID

Acid – $C_{24}H_{32}O_9 \cdot 5H_2O$ (anal. C, H), mp 188-190°, structure uncertain ⟨65-7⟩ ⟨67-10⟩.

17-β-D-GLUCOPYRANOSIDURONIC ACID-3-BENZYL ETHER-16-ACETATE

2′,3′,4′-Tri-*O*-acetyl methyl ester – $C_{40}H_{48}O_{13}$ (anal. C, H), mp 202-204°, $[\alpha]_D$ – 20° (ethanol) ⟨67-10⟩.

Allied Characterization Data

3-GLUCURONIDE

CC ⟨67-25⟩ • TLC ⟨67-25⟩.

16-GLUCURONIDE

CC ⟨67-29⟩ • PC ⟨67-25⟩ ⟨67-29⟩ • TLC ⟨67-25⟩.

17-GLUCURONIDE

CC, PC ⟨67-29⟩.

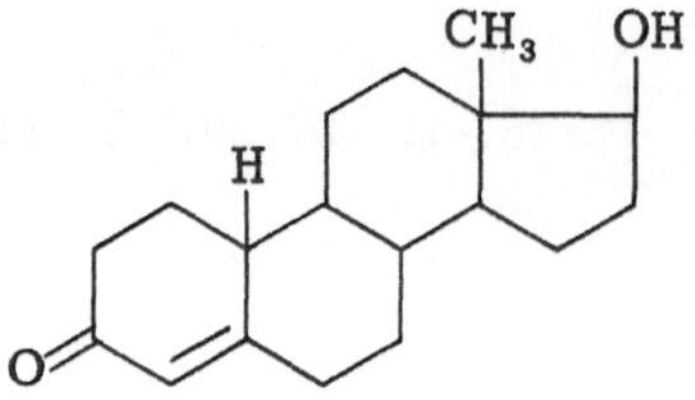

CH$_3$ OH

H

O

17β-Hydroxyestra-4-en-3-one
(19-Nortestosterone)

17-β-D-GLUCOPYRANOSIDURONIC ACID

Enzymatic Synthesis ⟨57-10⟩.

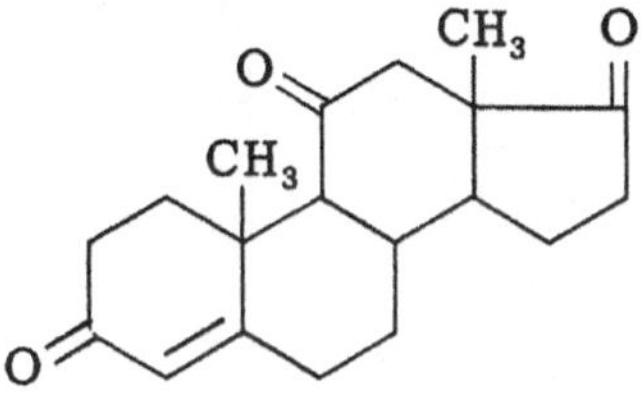

Androst-4-ene-3,11,17-trione
(Adrenosterone)

β-D-GLUCOPYRANOSIDURONIC ACID

Enzymatic synthesis ⟨57-10⟩.

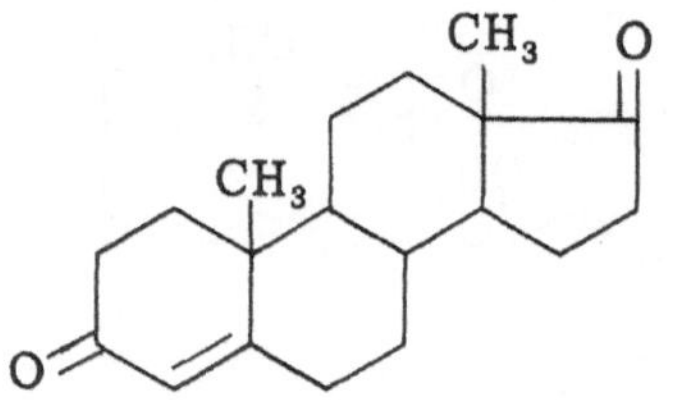

Androst-4-ene-3,17-dione
(Androstenedione)

3,5-DIEN-3-OL-β-D-GLUCOPYRANOSIDURONIC ACID

Acid – $C_{25}H_{34}O_8$ (anal. C, H), mp 176-178°, $[\alpha]_D^{23}$ – 119.1° (chloroform) ⟨59-7⟩ • mp 174-177°, λ_{max} 230 mμ ⟨62-15⟩ • PC, TLC ⟨67-19⟩.

2′,3′,4′-Tri-*O*-acetyl methyl ester – $C_{32}H_{42}O_{11} \cdot C_2H_5OH$ (anal. C, H), mp 199-201°, $[\alpha]_D^{23}$ + 119° ⟨59-7⟩ • IR ⟨59-8⟩ • mp 185-188° ⟨57-11⟩ • $\lambda_{max}^{methanol}$ 238 mμ ⟨58-9⟩.

3,5-DIEN-3-OL-SULFATE

Sodium salt – mp 194-200°, IR, TLC, CC ⟨67-23⟩.

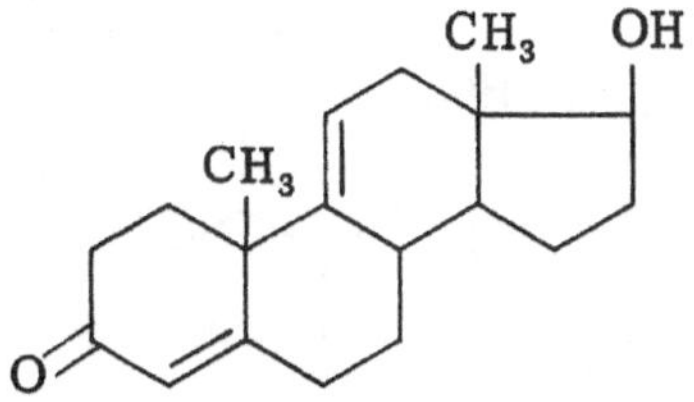

17β-Hydroxyandrosta-4,9(11)-dien-3-one
(9-Dehydrotestosterone)

17-β-D-GLUCOPYRANOSIDURONIC ACID

Enzymatic synthesis ⟨57-10⟩.

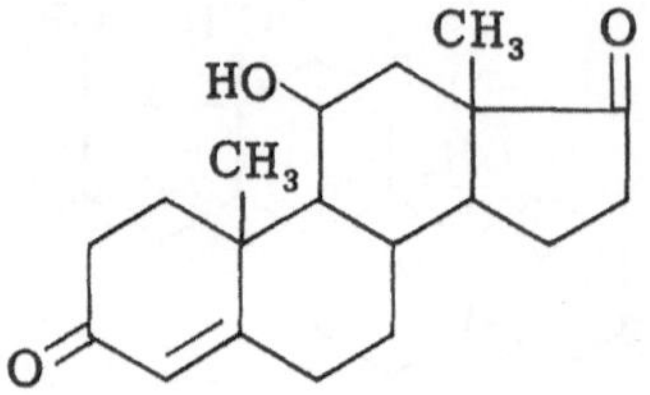

11β-Hydroxyandrost-4-ene-3,17-dione

3,5-DIEN-3-OL-β-D-GLUCOPYRANOSIDURONIC ACID

Acid — $\lambda_{max}^{abs.\ ethanol}$ 230 mμ ⟨63-20⟩.

β-D-GLUCOPYRANOSIDURONIC ACID

Enzymatic synthesis ⟨57-10⟩.

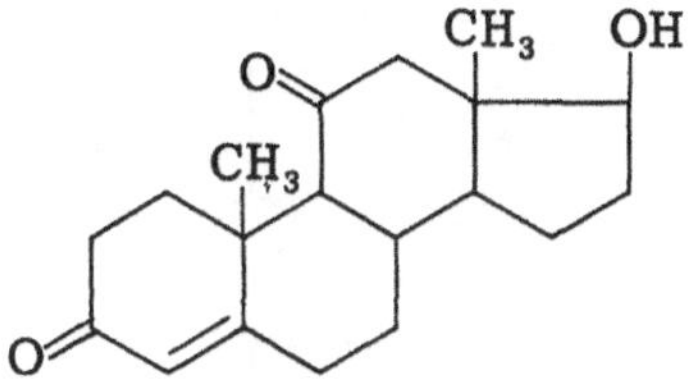

17β-Hydroxyandrost-4-ene-3,11-dione
(11-Ketotestosterone)

17-β-D-GLUCOPYRANOSIDURONIC ACID

Enzymatic synthesis ⟨57-10⟩.

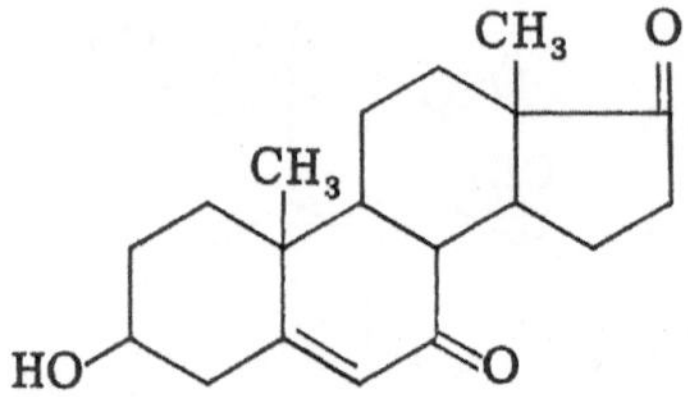

3β-Hydroxyandrost-5-ene-7,17-dione
(7-Ketodehydroisoandrosterone,
7-Ketodehydroepiandrosterone)

3-SULFATE

Sodium salt – $C_{19}H_{25}NaO_6S{\cdot}2H_2O$ (anal. C, H), mp 162-165°, IR ⟨61-1⟩⟨62-2⟩⟨65-11⟩.

Allied Characterization Data

3-SULFATE

CC ⟨62-1⟩ • PC ⟨62-1⟩⟨62-2⟩⟨66-32⟩.

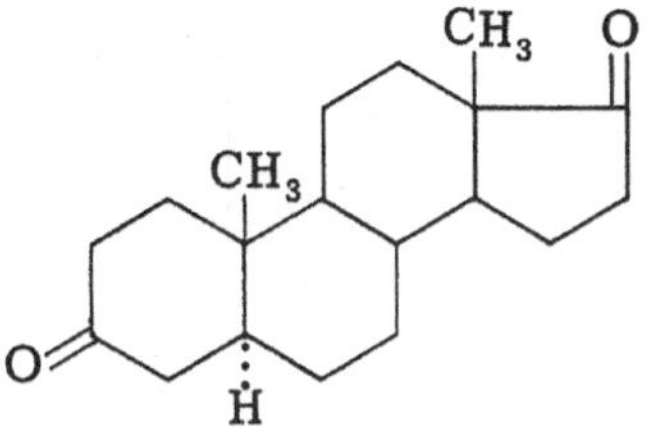

CH₃ O
CH₃
O
H

5α-Androstane-3,17-dione
(Androstanedione)

β-D-GLUCOPYRANOSIDURONIC ACID (Unspecified Location)

PC, TLC ⟨67-19⟩.

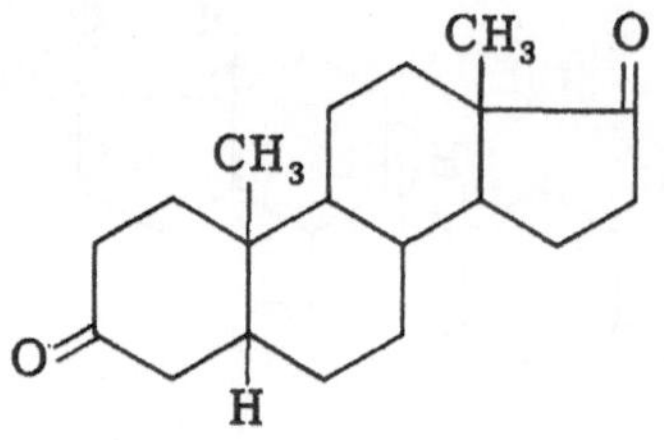

5β-Androstane-3,17-dione
(Etiocholanedione)

β-D-GLUCOPYRANOSIDURONIC ACID (Unspecified Location)

CC, TLC ⟨67-19⟩.

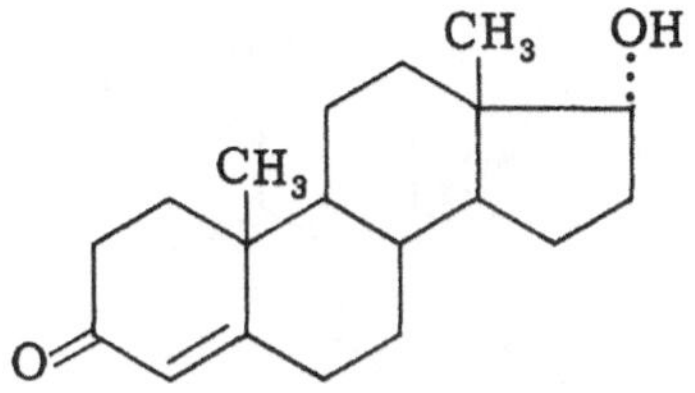

17α-Hydroxyandrost-4-en-3-one
(Epitestosterone)

17-β-D-GLUCOPYRANOSIDURONIC ACID

Enzymatic synthesis ⟨57-10⟩.

17-SULFATE

CC ⟨66-15⟩⟨67-8⟩ • IR ⟨67-8⟩ • PC ⟨66-15⟩.

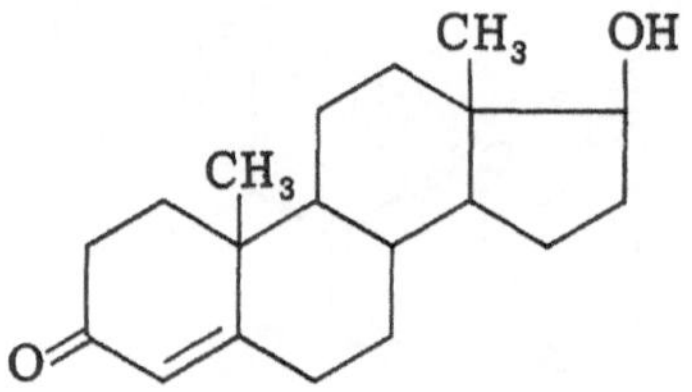

17β-Hydroxyandrost-4-en-3-one
(Testosterone)

17-α-D-GLUCOFURANOSIDURONIC ACID

Acid – mp 218-225°, IR ⟨63-16⟩.

γ-Lactone – $C_{25}H_{34}O_7$ (anal. C, H), mp 205-207°, $[\alpha]_D^{25}$ + 134° (dimethylformamide), IR ⟨63-16⟩.

17-β-D-GLUCOFURANOSIDURONIC ACID

Amide – $C_{25}H_{34}NO_7$, mp 186-194° ⟨63-16⟩.

γ-Lactone – $C_{25}H_{34}O_7$ (anal. C, H), mp 240-242°, $[\alpha]_D^{25}$ + 17° (dimethylformamide), IR ⟨63-16⟩.

17-β-D-GLUCOPYRANOSIDURONIC ACID

Acid – $C_{25}H_{36}O_8 \cdot H_2O$ (anal. C, H), mp 182-183.5°, $[\alpha]_D^{23}$ + 32° (ethanol) ⟨59-7⟩ • mp 177-181°, CCD ⟨64-8⟩ • $\lambda_{max}^{ethanol}$ 240 mμ, CCD, IR ORD ⟨66-22⟩ • 7α-^{3}H-^{14}C (glucuronide label) ⟨66-36⟩.

Methyl ester – $C_{26}H_{38}O_8$ (anal. C, H), mp 230.5-233.5°, mp 237-238°, CCD ⟨64-8⟩ • mp 230.5-233.5°, IR, 1,2-^{3}H ⟨66-22⟩.

Potassium salt – mp > 290° ⟨66-22⟩.

Sodium salt – $C_{25}H_{35}NaO_8 \cdot 3H_2O$ (anal. C, H, Na), mp 295-300°, $[\alpha]_D^{26}$ + 14° (water), $[\alpha]_{5460}^{26}$ + 24° (water), CC, PC, TLC ⟨64-5⟩⟨66-35⟩ • λ_{max} 240 mμ ⟨59-3⟩ • (anal. C, H, Na), 4-^{14}C ⟨64-20⟩.

2′,3′,4′-Tri-*O*-acetyl methyl ester – $C_{32}H_{44}O_{11}\cdot C_2H_5OH$ (anal. C, H), mp 186-189°, $[\alpha]_D^{20}$ + 28.3° (chloroform) ⟨39-3⟩ • (anal. C, H), mp 188.2-188.5°, $[\alpha]_D^{25.5}$ + 48° (chloroform) ⟨59-7⟩ • IR ⟨59-8⟩ • (anal. C, H), mp 188.2-188.5°, $\lambda_{max}^{methanol}$ 240 mμ, IR ⟨58-9⟩ • mp 186-188°, $[\alpha]_D^{26}$ + 20.5° (dioxane), $[\alpha]_{5460}^{26}$ + 27° (dioxane), $[\alpha]_{4360}^{26}$ + 39° (dioxane) ⟨64-5⟩ • mp 186-188° ⟨64-8⟩ • mp 186-188°, $[\alpha]_D^{26}$ + 20° (dioxane), TLC ⟨66-35⟩ • mp 186.5-187.7°, IR, NMR, ORD, 1,2-^{3}H ⟨66-22⟩ • 7α-^{3}H-C1′-^{14}C ⟨65-23⟩ • mp 179.5-181.5°, $\lambda_{max}^{methanol}$ 240 mμ ⟨59-3⟩.

Allied Characterization Data

GLUCURONIDE

PC ⟨62-1⟩⟨67-19⟩, TLC ⟨66-38⟩⟨67-19⟩

* * * * *

17-SULFATE

Ammonium salt – $C_{19}H_{31}NO_5S\cdot\frac{1}{4}H_2O$ (anal. C, H, N, S, H_2O), mp 201-203°, $[\alpha]_D^{25}$ + 70° (methanol), $\lambda_{max}^{methanol}$ 240 mμ (ε 15,000) ⟨66-25⟩ • TLC ⟨67-24⟩.

Potassium salt – $C_{19}H_{27}KO_5S$ (anal. C, H, S), no sharp mp, amorphous ⟨66-30⟩ • $C_{19}H_{27}KO_5S\cdot H_2O$ (anal. C, H, K, S, H_2O), mp 260°, $[\alpha]_D^{25}$ + 58° (methanol), $\lambda_{max}^{methanol}$ 240 mμ (ε 14,800) ⟨66-25⟩ • mp 266°, IR, PC ⟨62-2⟩ • IR ⟨65-11⟩.

Sodium salt – $C_{19}H_{27}NaO_5S$ (anal. C, H, Na), mp 215°, $[\alpha]_D^{25}$ + 74.5° (water), $[\alpha]_D^{25}$ + 68° (ethanol), λ_{max}^{water} 248-249 mμ (ε 20,300), $\lambda_{max}^{alcohol}$ 241 mμ (ε 17,700) ⟨49-2⟩ • $C_{19}H_{27}NaO_5S\cdot\frac{1}{2}H_2O$ (anal. C, H, Na, S, H_2O), mp 213-215°, $[\alpha]_D^{25}$ + 74° (water), $\lambda_{max}^{methanol}$ 240 mμ (ε 16,300) ⟨66-25⟩ • mp 210-213° ⟨57-2⟩.

Triethylammonium salt – $C_{25}H_{43}NO_5S$ (anal. C, H, N, S), mp 158-163°, $[\alpha]_D^{25}$ + 64° (chloroform), $\lambda_{max}^{methanol}$ 241 mμ (ε 16,400), NMR ⟨68-2⟩.

17-SULFATE-3-SEMICARBAZONE

Sodium salt – mp > 300°, dec. begins at 240° ⟨49-2⟩.

Allied Characterization Data

17-SULFATE

PC ⟨62-1⟩ ⟨62-2⟩ ⟨64-1⟩ ⟨67-30⟩ • PE ⟨61-2⟩ ⟨64-1⟩ • TLC ⟨67-24⟩.

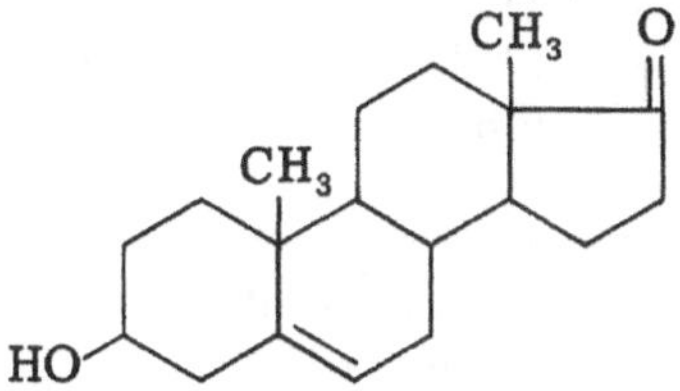

3β-Hydroxyandrost-5-en-17-one
(Dehydroisoandrosterone, Dehydroepiandrosterone, Androstenolone)

3-2′-AMINO-2′-DEOXY-β-D-GLUCOPYRANOSIDE

2′,3′,4′,6′-Tetraacetyl – $C_{33}H_{47}NO_{10}$, mp 260-261°, $[\alpha]_D$ + 8.2° (methanol) ⟨68-5⟩.

3-α-D-GLUCOFURANOSIDURONIC ACID

Amide – $C_{25}H_{37}NO_7$ (anal. N), mp 197-205°, IR ⟨63-16⟩.

γ-Lactone – $C_{25}H_{34}O_7$ (anal. C, H), mp 287-288°, $[\alpha]_D^{25}$ + 77° (dimethylformamide), IR ⟨63-16⟩.

3-β-D-GLUCOFURANOSIDURONIC ACID

Amide – $C_{25}H_{37}NO_7$ (anal. C, H, N), mp 175-177°, $[\alpha]_D^{25}$ – 78° (dimethylformamide), IR ⟨63-16⟩.

3-β-D-GLUCOPYRANOSIDURONIC ACID

Acid – $C_{25}H_{36}O_8$ (anal. C, H), mp 262-264° ⟨38-3⟩ ⟨39-3⟩ • mp 258-262° ⟨63-16⟩ • $C_{25}H_{36}O_8 \cdot CH_3OH$, mp 230-232°, mp 262-264°, $[\alpha]_D^{25}$ + 36° ⟨59-7⟩ • $C_{25}H_{36}O_8$ (anal. C, H), 170-210° ⟨59-3⟩ • mp 186-189° ⟨61-17⟩ • mp 258-261° ⟨61-4⟩.

γ-Lactone – $C_{25}H_{34}O_7$ (anal. C, H), mp 226-227°, $[\alpha]_D^{25}$ – 51° (dimethylformamide), IR ⟨63-16⟩.

Sodium salt – $C_{25}H_{35}NaO_8 \cdot 2H_2O$ (anal. C, H, Na), mp > 350°, $[\alpha]_D^{26.5}$ – 26° (water), $[\alpha]_{5460}^{26.5}$ – 46° (water), ⟨64-5⟩ • mp 200°, $[\alpha]_D^{12.8}$ – 9.8° (methanol), IR, TLC ⟨59-3⟩⟨61-17⟩.

2′,3′,4′-Tri-*O*-acetyl methyl ester – $C_{32}H_{44}O_4$ (anal. C, H), mp 193-196°, $[\alpha]_D^{20}$ – 19.7° (chloroform), $[\alpha]_D^{20}$ – 16.2° (benzene) ⟨39-3⟩ • mp 194-196°, $[\alpha]_D^{25}$ – 8.4° (chloroform) ⟨38-3⟩ • $C_{32}H_{44}O_{11} \cdot C_2H_5OH$ (anal. C, H), mp 195-195.5°, $[\alpha]_D^{23}$ – 11.3° (chloroform) ⟨59-7⟩ • IR ⟨59-8⟩ • (anal. C, H), mp 190-193° ⟨63-16⟩ • $C_{32}H_{44}O_{11}$ (anal. C, H, O), mp 186.5-188°, $[\alpha]_D^{20}$ – 12.7° (chloroform) ⟨59-3⟩ • mp 194-195°, $[\alpha]_D^{26.5}$ – 17° (dioxane), $[\alpha]_D^{26.5}$ 0 ± 4° (dioxane), $[\alpha]_D^{26.5}$ + 8° (dioxane), ⟨64-5⟩ • mp 194-195.5° ⟨61-4⟩.

Allied Characterization Data

3-GLUCURONIDE

CC ⟨62-8⟩⟨63-17⟩⟨66-41⟩ • CCD ⟨59-6⟩ • PC ⟨55-2⟩ ⟨59-6⟩⟨62-1⟩⟨62-8⟩⟨66-12⟩⟨67-19⟩ • PE ⟨64-11⟩ • TLC ⟨66-37⟩⟨67-19⟩.

3-GLUCURONIDE-TRI-*O*-TRIMETHYLSILYL METHYL ESTER

GLC ⟨67-31⟩.

* * * * *

3-SULFATE

Acid – ethyl acetate solution ⟨58-2⟩.

Ammonium salt – $C_{19}H_{31}NO_5S \cdot \frac{3}{4}H_2O$ (anal. C, H, N, S, H_2O), mp 206-207°, $[\alpha]_D^{25}$ + 19° ⟨66-25⟩ • enzymatic synthesis 7α-^{3}H-^{35}S ⟨64-22⟩ • mp 202-204° ⟨56-1⟩ • IR ⟨63-18⟩⟨64-21⟩.

N-(β-Bromoethyl)-N,N-dimethylamine salt – mp 166-170°, $[\alpha]_D^{20}$ + 7.6° (methanol) ⟨56-2⟩.

Licocaine salt – mp 173-176°, $[\alpha]_D^{20}$ + 6.9° (methanol) ⟨56-2⟩.

Methyl ester – $C_{20}H_{30}O_5S$ (anal. C, H, S), mp 128-131°, $[\alpha]_D^{23.5}$ −1° (chloroform), IR ⟨60-6⟩.

Potassium salt – $C_{19}H_{27}KO_5S$ (hydrate?) (anal. K, S), mp 221-223° ⟨48-3⟩ • $C_{19}H_{27}KO_5S$ (anal. K), mp 219-223° ⟨58-1⟩⟨58-2⟩ • (anal. S), mp 218-220°, $[\alpha]_D^{19}$ + 6.9° (propylene glycol-ethanol 1:3) ⟨53-3⟩ • IR ⟨62-2⟩ ⟨65-11⟩.

Procaine salt – mp 179-180.5°, $[\alpha]_D^{20}$ + 8.5° (methanol) ⟨56-2⟩.

Pyridine salt – mp 194-195° ⟨43-1⟩ • mp 193-194° ⟨56-1⟩ • IR ⟨64-21⟩.

Sodium salt – $C_{19}H_{27}NaO_5S$ (anal. Na) mp 192-193° ⟨43-1⟩ • mp 184-187°, $[\alpha]_D^{20}$ + 10.7° (methanol) ⟨56-1⟩ • mp 183-187° ⟨57-2⟩ • $C_{19}H_{27}NaO_5S \cdot 2H_2O$ (anal. C, H), mp 172-174° ⟨60-6⟩ • mp 172-174°, IR ⟨62-2⟩ • IR ⟨64-21⟩⟨65-11⟩ • PC ⟨55-2⟩.

Triethylammonium salt – $C_{25}H_{43}NO_5S$ (anal. C, H, N, S), mp 220-222°, $[\alpha]_D^{25}$ + 1° (chloroform), NMR ⟨68-2⟩.

3-SULFATE-17-OXIME

Ammonium salt – mp > 205° ⟨66-3⟩.

Sodium salt – mp 150-152°, $[\alpha]_D^{20}$ − 26.5° (methanol) ⟨56-2⟩.

3-SULFATE-1-GLYCERYL ESTER

Dipalmitoyl – $C_{54}H_{90}O_9S$ (anal. C, H), mp 65-67°, CC, PC ⟨61-12⟩.

Allied Characterization Data

3-SULFATE

CC ⟨57-3⟩ ⟨61-6⟩ ⟨62-1⟩ ⟨62-8⟩ ⟨63-3⟩ ⟨63-5⟩ ⟨63-18⟩ ⟨64-21⟩ ⟨66-4⟩ ⟨66-5⟩ ⟨66-27⟩ ⟨66-40⟩ ⟨66-41⟩ ⟨67-23⟩ ⟨67-26⟩ • CCD ⟨59-6⟩ ⟨63-5⟩ • PC ⟨55-2⟩ ⟨56-4⟩ ⟨56-5⟩ ⟨58-3⟩ ⟨59-6⟩ ⟨60-2⟩ ⟨60-15⟩ ⟨61-6⟩ ⟨62-1⟩ ⟨62-8⟩ ⟨63-5⟩ ⟨64-1⟩ ⟨66-32⟩ ⟨67-23⟩ ⟨67-26⟩ • PE ⟨57-2⟩ ⟨57-3⟩ ⟨58-4⟩ ⟨59-5⟩ ⟨61-2⟩ ⟨62-2⟩ ⟨63-5⟩ ⟨64-1⟩ ⟨64-11⟩ • TLC ⟨64-3⟩ ⟨66-21⟩ ⟨66-27⟩ ⟨66-37⟩ ⟨67-23⟩ ⟨67-24⟩.

3-SULFATE-17-OXIME

CC ⟨63-3⟩ ⟨64-21⟩ • IR ⟨64-21⟩.

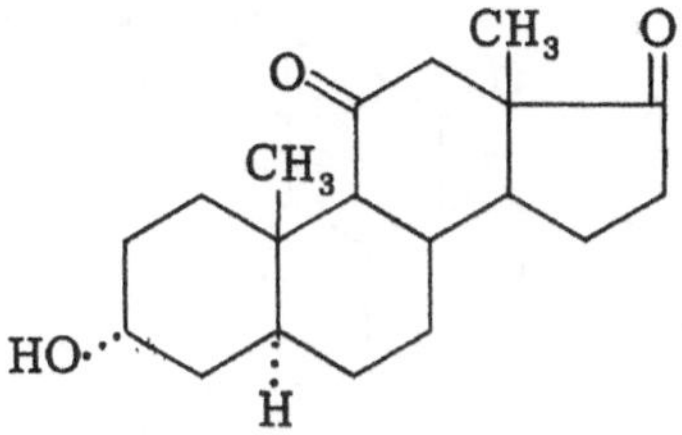

3 α-Hydroxy-5 α-androstan-11,17-dione
(11-Ketoandrosterone)

3-GLUCURONIDE

TLC ⟨66-43⟩.

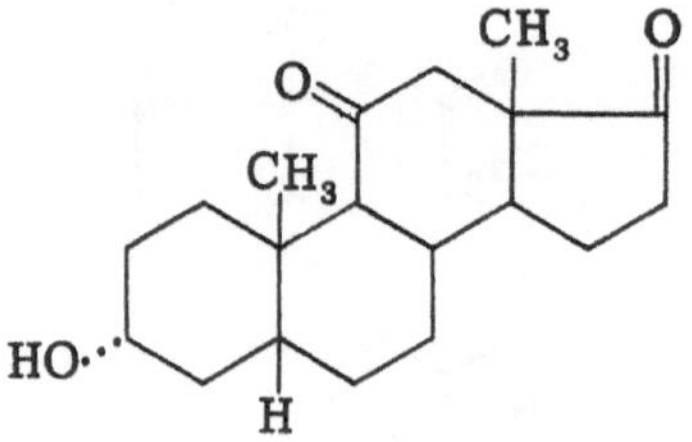

3α-Hydroxy-5β-androstane-11,17-dione
(11-Ketoetiocholanolone)

3-β-D-GLUCOPYRANOSIDURONIC ACID

2′,3′,4′-Tri-*O*-acetyl methyl ester – $C_{32}H_{44}O_{12}$ (anal. C, H), mp 168.5-170°, $[\alpha]_D^{20}$ + 58° (chloroform), CC, IR, PC ⟨64-6⟩.

Allied Characterization Data

3-GLUCURONIDE

CC ⟨58-8⟩ ⟨60-8⟩ • PC ⟨58-8⟩ • TLC ⟨66-43⟩.

3-GLUCURONIDE TRI-*O*-TRIMETHYLSILYL METHYL ESTER

GLC, TLC, MS ⟨67-16⟩.

* * * * *

3-SULFATE

Sodium salt – mp 160-162°, IR, PC ⟨62-2⟩ ⟨65-11⟩.

Allied Characterization Data

3-SULFATE

PC ⟨62-1⟩.

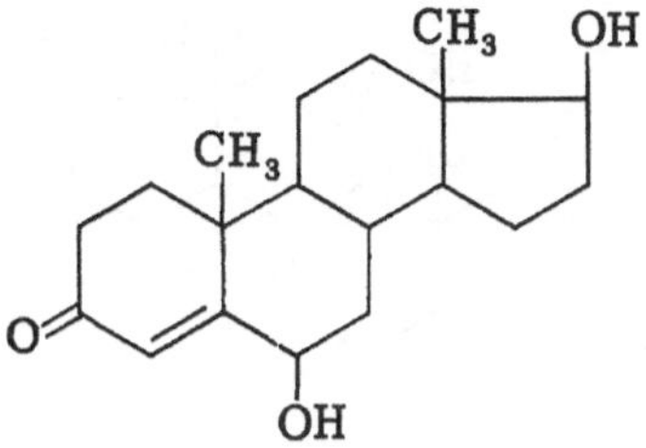

$6\beta,17\beta$-Dihydroxyandrost-4-en-3-one
(6β-Hydroxytestosterone)

β-D-GLUCOPYRANOSIDURONIC ACID (Unspecified Location)

Enzymatic synthesis ⟨57-10⟩.

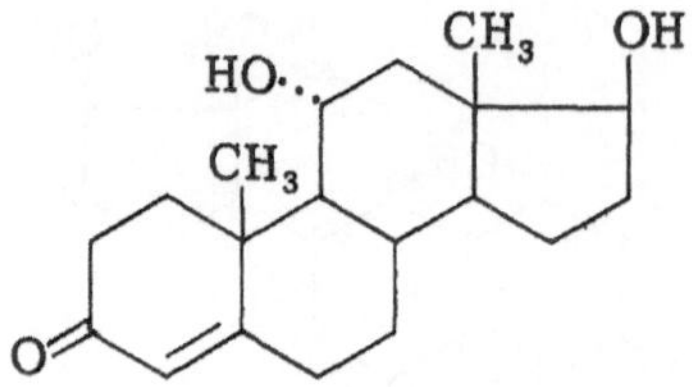

11α,17β-Dihydroxyandrost-4-en-3-one
(11α-Hydroxytestosterone)

β-D-GLUCOPYRANOSIDURONIC ACID (Unspecified Location)

Enzymatic synthesis ⟨57-10⟩.

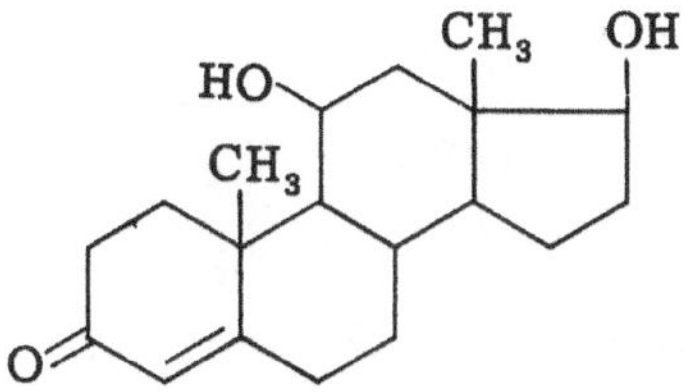

11β,17β-Dihydroxyandrost-4-en-3-one
(11β-Hydroxytestosterone)

β-D-GLUCOPYRANOSIDURONIC ACID (Unspecified Location)

Enzymatic synthesis ⟨57-10⟩.

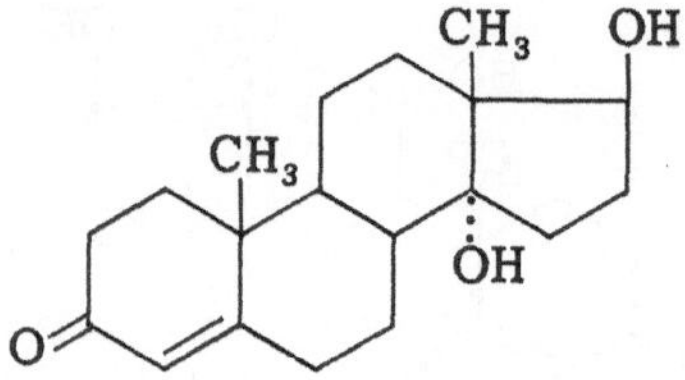

14α,17β-Dihydroxyandrost-4-en-3-one
(14α-Hydroxytestosterone)

β-D-GLUCOPYRANOSIDURONIC ACID (Unspecified Location)

Enzymatic synthesis ⟨57-10⟩.

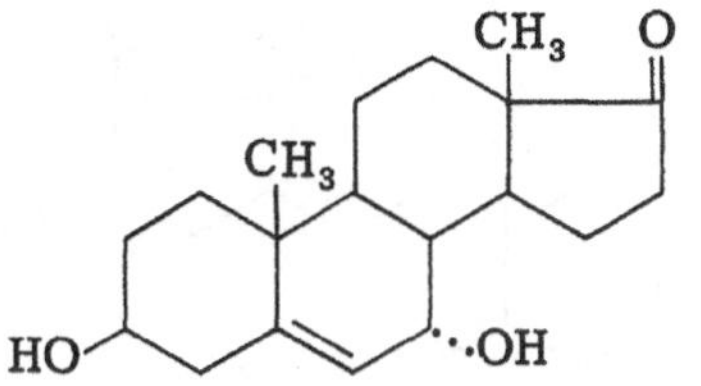

3β,7α-Dihydroxyandrost-5-en-17-one
(7α-Hydroxydehydroisoandrosterone,
7α-Hydroxydehydroepiandrosterone)

3-SULFATE

Sodium salt – mp 217° ⟨66-42⟩.

Allied Characterization Data

3-SULFATE

PC ⟨62-12⟩⟨66-10⟩ • PE ⟨62-12⟩.

3,17-DISULFATE

PE ⟨62-12⟩.

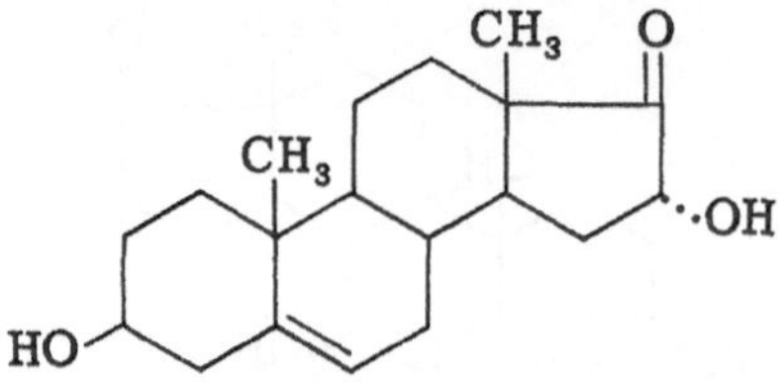

3β,16α-Dihydroxyandrost-5-en-17-one
(16α-Hydroxydehydroisoandrosterone,
16α-Hydroxydehydroepiandrosterone)

β-D-GLUCOPYRANOSIDURONIC ACID (Unspecified Location)

PC, TLC ⟨67-19⟩.

3-SULFATE

Ammonium salt – ⟨65-1⟩ • IR, 7α-^{3}H ⟨67-33⟩.

Potassium salt – CC, IR ⟨64-21⟩.

Allied Characterization Data

3-SULFATE

CC ⟨65-1⟩ ⟨66-5⟩ ⟨67-33⟩ ⟨67-34⟩ • PC ⟨65-1⟩ ⟨66-16⟩ ⟨67-33⟩ ⟨67-34⟩ • TLC ⟨66-16⟩.

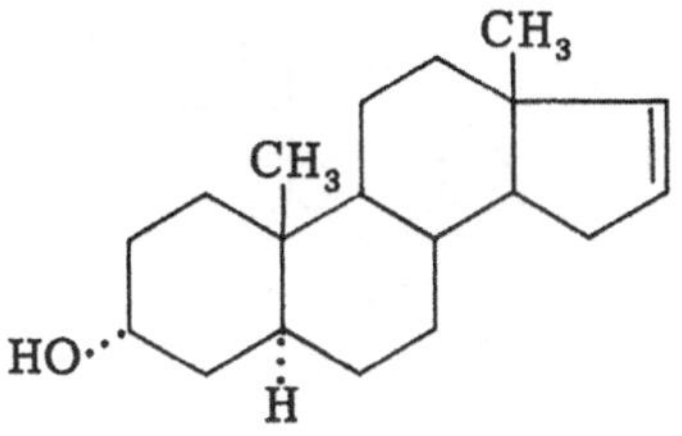

5α-Androst-16-en-3α-ol

3-β-D-GLUCOPYRANOSIDURONIC ACID

Acid – $C_{25}H_{38}O_7$ (anal. C, H), mp 169-174° ⟨60-4⟩ ⟨61-3⟩.

2′,3′,4′-Tri-*O*-acetyl methyl ester – $C_{32}H_{46}O_{10}$ (anal. C, H), mp 171-172° ⟨60-4⟩ ⟨61-3⟩.

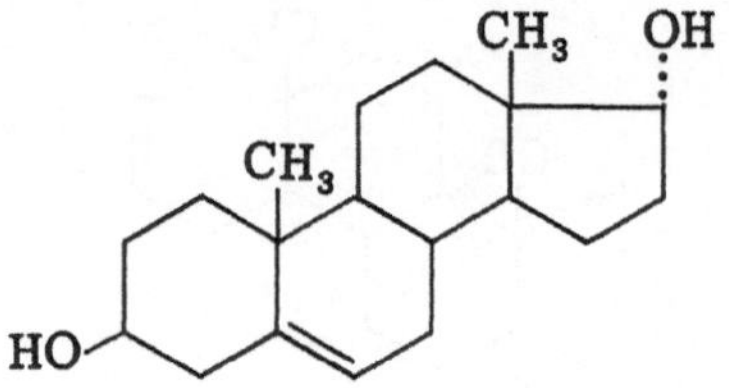

Androst-5-ene-3β,17α-diol

3-SULFATE

CC ⟨66-10⟩.

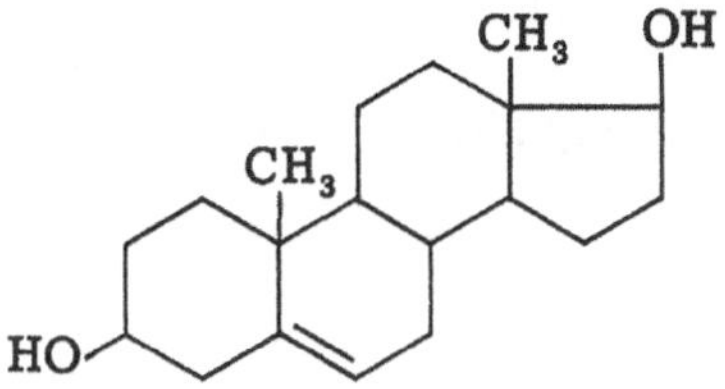

Androst-5-ene-3β,17β-diol
(Androstenediol)

17-2′-AMINO-2′-DEOXY-β-D-GLUCOPYRANOSIDE-3-ACETATE

2′,3′,4′,6′-Tetraacetyl – $C_{35}H_{52}NO_{11}$, mp 240-242°, $[\alpha]_D$ – 31.6° (methanol) ⟨68-5⟩.

3,17-DI-β-D-GLUCOPYRANOSIDURONIC ACID

Di-2′,3′,4′-Tri-*O*-acetyl methyl ester – $C_{45}H_{62}O_{20}$ (anal. C, H), mp 271-272°, $[\alpha]_D^{23}$ – 53° (dioxane) ⟨66-35⟩.

Allied Characterization Data

GLUCURONIDE

CC ⟨67-19⟩ ⟨67-26⟩ • PC ⟨67-19⟩ ⟨67-26⟩ • TLC ⟨67-19⟩.

* * * * *

3-SULFATE

Ammonium salt – mp 206-209° ⟨66-3⟩ ⟨66-4⟩ • ^3H–^{35}S ⟨63-1⟩ • mp 210-213°, IR ⟨64-21⟩.

Potassium salt – mp 260-265°, IR, PC ⟨62-2⟩ ⟨65-1⟩ • mp 254-255° ⟨57-2⟩ • IR ⟨65-11⟩.

Pyridine salt – CC ⟨66-5⟩.

3-SULFATE-17-ACETATE

Potassium salt – mp 250° ⟨65-1⟩.

3,17-DISULFATE

Diammonium salt – mp 249-250°, $[\alpha]_D^{25}$ – 38° (methanol) ⟨68-3⟩.

Dipotassium salt – mp 240-245° ⟨57-2⟩.

Disodium salt – mp 260°, IR ⟨62-2⟩ ⟨65-11⟩.

Allied Characterization Data

3-SULFATE

PE ⟨62-2⟩ • TLC ⟨67-24⟩.

MONOSULFATE (Unspecified Location)

PC, PE ⟨62-14⟩, ⟨63-19⟩ • TLC ⟨64-3⟩.

DISULFATE

PC ⟨62-2⟩ ⟨62-14⟩, ⟨63-19⟩ • PE ⟨62-2⟩ ⟨62-14⟩ ⟨63-19⟩.

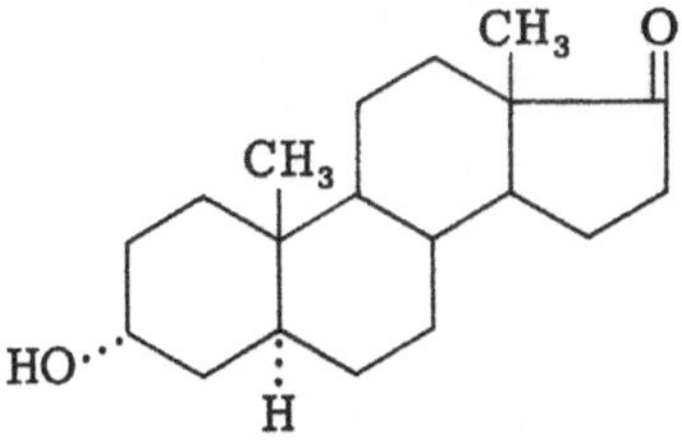

3α-Hydroxy-5α-androstan-17-one
(Androsterone)

3-2′-AMINO-2′-DEOXY-β-D-GLUCOPYRANOSIDE

2′-Acetyl – $C_{27}H_{43}NO_7 \cdot 2H_2O$, mp 154-159°, $[\alpha]_D$ + 34.2° (water) ⟨68-5⟩.

2′,3′,4′,6′-Tetraacetyl – $C_{33}H_{49}NO_{10}$, mp 229-231°, $[\alpha]_D$ + 24.3° (chloroform) ⟨68-5⟩.

3-α-D-GLUCOFURANOSIDURONIC ACID

γ-Lactone – $C_{25}H_{36}O_7$, mp 192-198°, $[\alpha]_D^{25}$ + 107° (dimethylformamide), IR ⟨63-16⟩.

Potassium salt – mp 250°, IR ⟨63-16⟩.

3-β-D-GLUCOPYRANOSIDURONIC ACID

Acid – $C_{25}H_{38}O_8$ (anal. C, H), mp 148.5-149.5°, $[\alpha]_D^{21.5}$ + 20° (chloroform) ⟨64-6⟩ • (anal. C, H, O), mp 193-195°, IR, PC ⟨65-2⟩ • mp 168-170°, IR, TLC, PC ⟨61-17⟩ • isolated from urine ⟨60-10⟩ • IR, TLC ⟨66-43⟩ • (anal. C, H) ⟨60-5⟩ • noncrystalline, CC, CCD, PC ⟨63-7⟩.

Amide – $C_{25}H_{39}NO_7$, mp 204-209° ⟨63-16⟩.

Barium salt – $C_{25}H_{37}Ba_{1/2}O_8$, mp 230°, TLC ⟨61-17⟩.

Cinchonine salt – $(C_{25}H_{38}O_8)_3 \cdot C_{19}H_{22}N_2O$ (anal. N), mp 150-160°, IR, TLC ⟨61-17⟩.

Sodium salt — $C_{25}H_{37}NaO_8$, mp 205°, $[\alpha]_D^{20}$ + 38.5°, IR, TLC ⟨61-17⟩ • mp 211-215° ⟨60-17⟩ • $C_{25}H_{37}NaO_8 \cdot 3H_2O$ (anal. C, H, Na), mp 254-256°, $[\alpha]_D^{26}$ + 29° (water), $[\alpha]_{5460}^{26}$ + 33° (water) ⟨64-5⟩.

2′,3′,4′-Tri-*O*-acetyl methyl ester — $C_{32}H_{44}O_{11}$ (anal. C, H), mp 175.5-176°, $[\alpha]_D^{21.5}$ + 12.2° (chloroform), CC, IR, PC ⟨64-6⟩ • $C_{32}H_{46}O_{11}$ (anal. C, H), mp 176-178°, $[\alpha]_{5890}^{26}$ + 25° (dioxane), $[\alpha]_{5460}^{26}$ + 36° (dioxane) ⟨64-5⟩ • $C_{32}H_{46}O_{11} \cdot C_6H_6$ (anal. C, H), mp 105-110° ⟨64-5⟩ • $[\alpha]_D^{20}$ + 10.8° (chloroform), CCD ⟨63-7⟩.

Allied Characterization Data

3-GLUCURONIDE

CC ⟨57-4⟩⟨57-26⟩⟨58-8⟩⟨62-8⟩⟨64-6⟩⟨66-41⟩⟨66-43⟩⟨67-26⟩ • IR ⟨66-43⟩ • PC ⟨58-8⟩⟨62-1⟩⟨62-2⟩⟨62-8⟩⟨66-43⟩⟨67-19⟩⟨67-26⟩ • PE ⟨62-2⟩ • TLC ⟨66-37⟩⟨66-43⟩⟨67-19⟩.

3-GLUCURONIDE TRI-*O*-TRIMETHYLSILYL METHYL ESTER

GLC ⟨67-14⟩⟨67-16⟩⟨67-31⟩.

* * * * *

3-SULFATE

Acid — ethyl acetate solution ⟨58-2⟩.

Methyl ester — $C_{20}H_{32}O_5S$ (anal. C, H, S), mp 114-116°, $[\alpha]_D$ + 76° (chloroform), IR ⟨60-6⟩.

Potassium salt — $C_{19}H_{29}KO_5S$ (anal. K), mp 184-185° ⟨58-2⟩.

Pyridine salt — CC ⟨66-5⟩.

Sodium salt — $C_{19}H_{29}NaO_5S$, mp 212-214°, $[\alpha]_D^{23}$ + 78° (water),

IR ⟨65-19⟩ • (anal. C, H, Na), mp 144°, ⟨42-3⟩ • $C_{19}H_{29}NaO_5S \cdot H_2O$ (anal. C, H, Na, S), mp 190°, ⟨42-3⟩ • $C_{19}H_{29}NaO_5S \cdot 2H_2O$ (anal. C, H, S) mp 139-140° (resolidification), IR, PC ⟨62-2⟩ • (anal. C, H, S), mp 139-140° ⟨60-6⟩ • mp 140-142° ⟨57-2⟩ • PC ⟨55-2⟩ • IR ⟨65-11⟩.

Triethylammonium salt – $C_{25}H_{45}NO_5S$ (anal. C, H, N, S), mp 206-207°, $[\alpha]_D^{25}$ + 53° (chloroform), NMR ⟨68-2⟩.

3-SULFATE-17-SEMICARBAZONE

Sodium salt – mp 245° ⟨42-3⟩.

Allied Characterization Data

3-SULFATE

CC ⟨57-4⟩⟨62-8⟩⟨66-5⟩⟨67-26⟩ • PC ⟨56-5⟩⟨58-3⟩⟨60-3⟩⟨61-6⟩⟨62-1⟩⟨62-2⟩⟨62-8⟩⟨64-1⟩⟨66-32⟩⟨67-26⟩ • PE ⟨59-5⟩⟨61-2⟩⟨62-9⟩⟨64-1⟩ • TLC ⟨64-3⟩⟨66-37⟩.

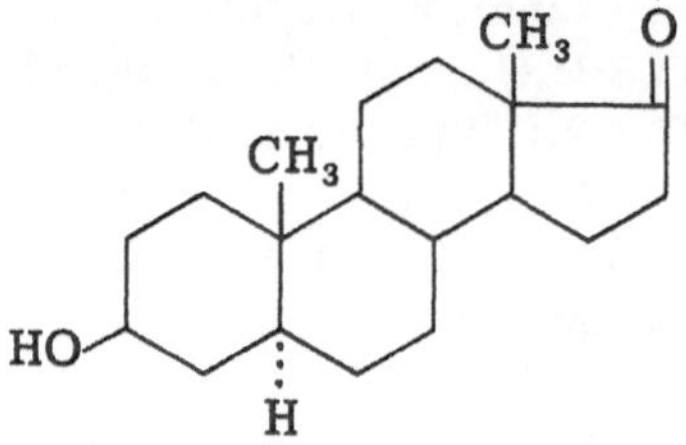

3β-Hydroxy-5α-androstan-17-one
(Isoandrosterone, Epiandrosterone)

3-2′-AMINO-2′-DEOXY-β-D-GLUCOPYRANOSIDE

2′,3′,4′,6′-Tetraacetyl – $C_{33}H_{49}NO_{10}$, mp 282-283°, $[\alpha]_D$ + 34.7° (methanol) ⟨68-5⟩.

3-β-D-GLUCOPYRANOSIDURONIC ACID

Acid – $C_{25}H_{38}O_8$ (anal. C, H, O), mp 196-199°, IR, PC ⟨65-2⟩ • (anal. C, H, O), PC ⟨59-3⟩.

Sodium salt – $C_{25}H_{37}NaO_8{\cdot}H_2O$ (anal. C, H, Na), mp 305-310°, $[\alpha]_D^{26}$ + 6° (water), $[\alpha]_{5460}^{26}$ + 12° (water) ⟨64-5⟩.

2′,3′,4′-Tri-*O*-acetyl methyl ester – $C_{32}H_{46}O_{11}$ (anal. C, H, O), mp 175-178°, $[\alpha]_D^{25.5}$ – 5.5° (dioxane), $[\alpha]_{5460}^{25.5}$ – 5° (dioxane) ⟨64-5⟩ • (anal. C, H), mp 163-164°, $[\alpha]_D^{20}$ + 19.3° (chloroform) ⟨59-3⟩ • (anal. C, H), mp 168-170°, $[\alpha]_D^{22}$ + 28° (chloroform) ⟨59-7⟩ • IR ⟨59-8⟩.

Allied Characterization Data

3-GLUCURONIDE

CC ⟨57-4⟩ ⟨67-26⟩ • PC ⟨62-1⟩ ⟨62-2⟩ ⟨66-12⟩ ⟨67-26⟩ • PE ⟨64-11⟩.

* * * * *

3-SULFATE

Acid – ethyl acetate solution ⟨58-2⟩.

Methyl ester – $C_{20}H_{32}O_5S$ (anal. C, H, S), mp 129-131°, $[\alpha]_D^{23.5}$ + 62° (chloroform), IR ⟨60-6⟩.

Potassium salt – $C_{19}H_{29}KO_5S$ (anal. K), mp 223-223.5° ⟨58-2⟩.

Sodium salt – $C_{19}H_{29}NaO_5S{\cdot}2H_2O$ (anal. C, H), mp 174-176° ⟨60-6⟩ • (anal. C, H), mp 174-176°, IR ⟨62-2⟩ • IR ⟨65-11⟩ • CC ⟨62-1⟩.

Triethylammonium salt – $C_{25}H_{45}NO_5S$ (anal. C, H, N, S), mp 210-212°, $[\alpha]_D^{25}$ + 52° (chloroform), NMR ⟨68-2⟩.

Allied Characterization Data

3-SULFATE

CC ⟨57-4⟩⟨65-1⟩⟨67-26⟩ • CCD ⟨59-6⟩ • PC ⟨56-5⟩ ⟨61-6⟩⟨62-1⟩⟨62-8⟩⟨64-1⟩ • PE ⟨62-2⟩⟨64-1⟩⟨64-11⟩ • TLC ⟨64-3⟩.

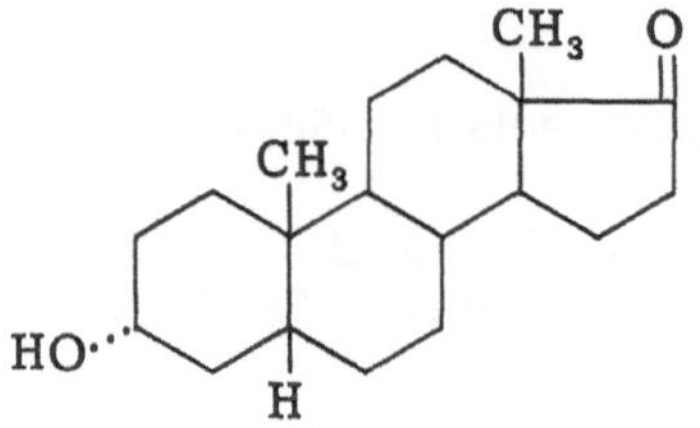

3α-Hydroxy-5β-androstan-17-one
(Etiocholanolone)

3-2′-AMINO-2′-DEOXY-β-D-GLUCOPYRANOSIDE

2′,3′,4′,6′-Tetraacetyl – $C_{33}H_{49}NO_{10}$, mp 179-181°, $[\alpha]_D$ + 42.6° (methanol) ⟨68-5⟩.

3-β-D-GLUCOPYRANOSIDURONIC ACID

Acid – $C_{25}H_{38}O_8$ (anal. C, H, O), mp 196-202°, $[M]_D$ 2408°, IR, CC, PC ⟨65-2⟩ • $C_{25}H_{38}O_8$ (anal. C, H), mp 175-177°, $[\alpha]_D^{21.5}$ + 33° (ethanol), CC, PC ⟨64-6⟩ • mp 204-206°, $[\alpha]_D^{20}$ + 26° ⟨63-7⟩ • IR, TLC ⟨66-43⟩.

Sodium salt – $C_{25}H_{37}NaO_8 \cdot 3H_2O$ (anal. C, H, Na), mp 220-230°, $[\alpha]_D^{26}$ + 42° (water), $[\alpha]_{5460}^{26}$ + 40° (water) ⟨64-5⟩ • $C_{25}H_{37}NaO_8 \cdot 4H_2O$ (anal. C, H), mp 227-230°, IR, PC, PE ⟨60-1⟩⟨63-7⟩ • mp 221°, PE ⟨60-17⟩.

2′,3′,4′-Tri-*O*-acetyl methyl ester – $C_{32}H_{46}O_{11}$ (anal. C, H, O), mp 160-162°, $[\alpha]_D^{26}$ + 27° (dioxane), $[\alpha]_{5460}^{26}$ + 35° (dioxane) ⟨64-5⟩ • (anal. C, H), mp 176.5-178°, $[\alpha]_D^{21.5}$ + 29° (chloroform), CC, PC ⟨64-6⟩ • mp 291°, $[\alpha]_D^{20}$ + 31° (chloroform), CCD ⟨63-7⟩ • CC, PC ⟨65-2⟩.

Allied Characterization Data

3-GLUCURONIDE

CC ⟨58-8⟩⟨59-6⟩⟨60-8⟩⟨66-43⟩⟨67-26⟩ • CCD ⟨59-6⟩⟨63-7⟩ • PE ⟨60-17⟩ • PC ⟨58-8⟩⟨60-17⟩⟨62-1⟩ ⟨67-19⟩ ⟨67-26⟩ • TLC ⟨66-43⟩⟨67-19⟩ • IR ⟨66-43⟩.

3-GLUCURONIDE-TRI-*O*-TRIMETHYLSILYL METHYL ESTER

GLC, TLC ⟨67-16⟩.

*　*　*　*　*

3-SULFATE

Methyl ester – $C_{20}H_{32}O_5S$ (anal. C, H, S), mp 126-128°, $[\alpha]_D^{23.5}$ + 97° (chloroform), IR ⟨60-6⟩.

Sodium salt – $C_{19}H_{29}NaO_5S \cdot 2H_2O$ (anal. C, H), mp 158-159° ⟨60-6⟩⟨62-2⟩ • IR ⟨62-2⟩⟨65-11⟩ • PC ⟨62-2⟩.

Triethylammonium salt – $C_{25}H_{45}NO_5S$ (anal. C, H, N, S), mp 171-173°, $[\alpha]_D^{25}$ + 62° (chloroform), NMR ⟨68-2⟩.

Allied Characterization Data

3-SULFATE

CC ⟨66-5⟩⟨67-26⟩ • PE ⟨64-1⟩ • PC ⟨60-3⟩⟨62-1⟩ ⟨64-1⟩⟨66-32⟩⟨67-26⟩ • TLC ⟨66-37⟩.

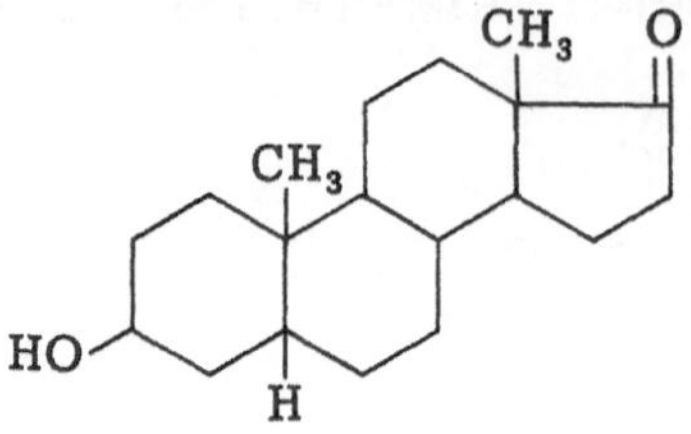

3β-Hydroxy-5β-androstan-17-one

3-β-D-GLUCOPYRANOSIDURONIC ACID

Acid – $C_{25}H_{38}O_8$ (anal. C, H, O), mp 198-200°, $[M]_D$ + 1973, CC, IR, PC ⟨65-2⟩.

2′,3′,4′-Tri-*O*-acetyl methyl ester – CC, PC ⟨65-2⟩.

Allied Characterization Data

3-GLUCURONIDE

CC, PC ⟨67-26⟩.

* * * * *

3-SULFATE

Methyl ester – $C_{20}H_{32}O_5S$ (anal. C, H, S), mp 106-108.5°, $[\alpha]_D^{23.5}$ + 73° (chloroform), IR ⟨60-6⟩.

Sodium salt – $C_{19}H_{29}NaO_5S \cdot 2H_2O$ (anal. C, H), mp 148-150° ⟨60-6⟩⟨62-2⟩ • mp 148-151°, IR, PC ⟨60-3⟩ • CC ⟨60-3⟩ • IR ⟨62-2⟩⟨65-11⟩ • PC ⟨62-1⟩⟨62-2⟩.

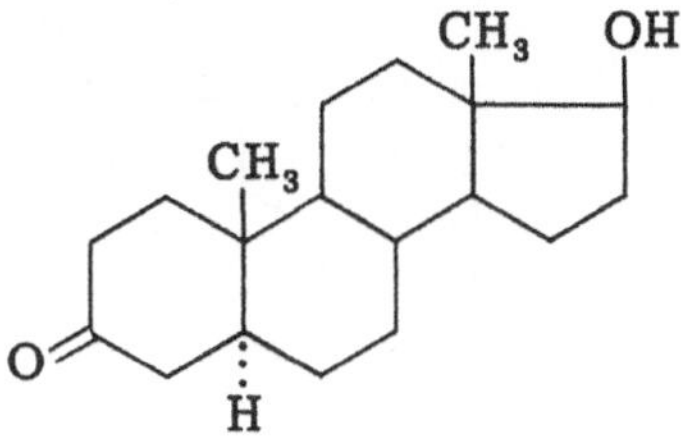

17β-Hydroxy-5α-androstan-3-one

17-2′-AMINO-2′-DEOXY-β-D-GLUCOPYRANOSIDE

2′,3′,4′,6′,-Tetraacetyl – $C_{33}H_{49}NO_{10}$, mp 123-126°, $[\alpha]_D$ + 15.1° (methanol) ⟨68-5⟩.

17-β-D-GLUCOPYRANOSIDURONIC ACID

4-^{14}C, TLC ⟨65-24⟩.

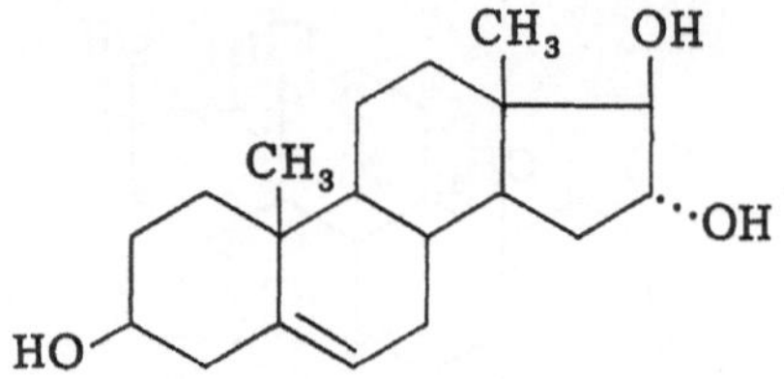

Androst-5-ene-3β,16α,17β-triol
(Androstenetriol)

GLUCURONIDE

PC, TLC ⟨67-19⟩.

3-SULFATE

CC, PC ⟨65-1⟩.

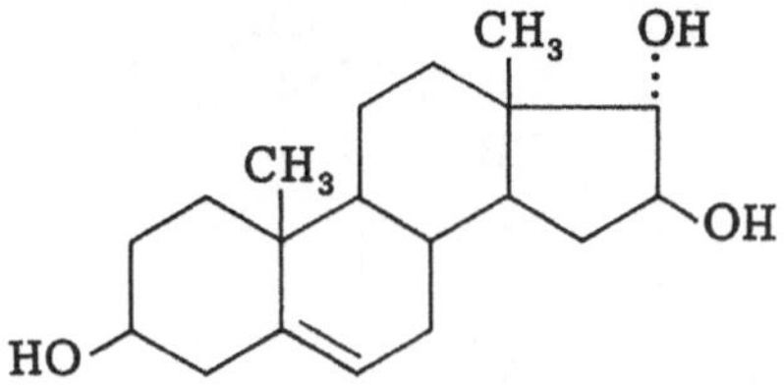

Androst-5-ene-3β,16β,17α-triol

SULFATE (Unspecified Location or Number)

Enzymatic synthesis, PC ⟨56-5⟩.

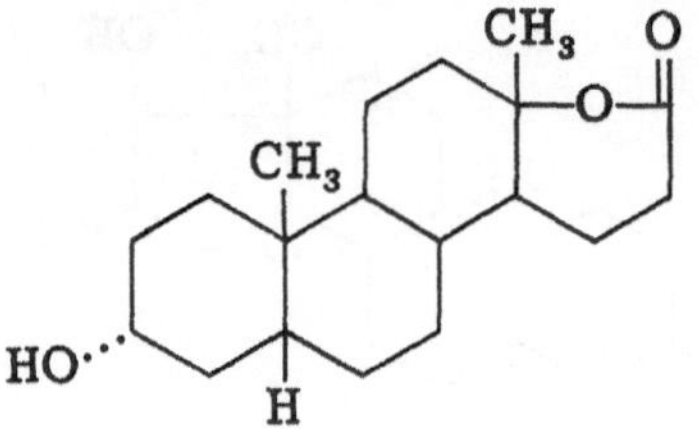

3α-Hydroxy-17a-oxa-D-homo-5β-androstan-17-one

GLUCURONIDE

⟨66-39⟩.

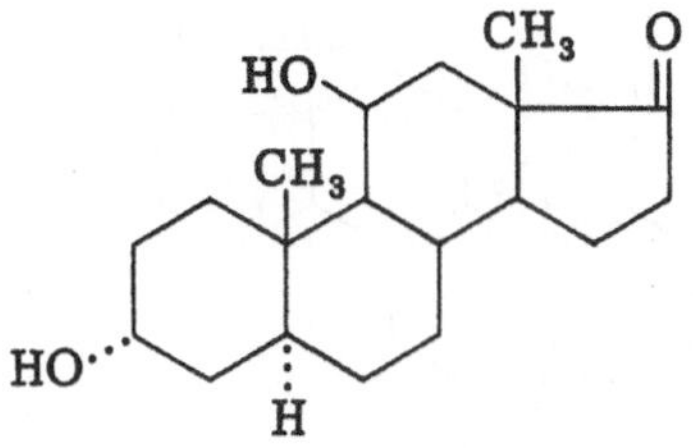

3α,11β-Dihydroxy-5α-androstan-17-one
(11β-Hydroxyandrosterone)

3-GLUCURONIDE

CC ⟨60-8⟩ ⟨64-6⟩ ⟨66-43⟩ • PC ⟨64-6⟩ • TLC ⟨66-43⟩.

3-GLUCURONIDE TRI-*O*-ACETYL METHYL ESTER

CC, PC ⟨64-6⟩.

3,11-DISULFATE

CC, PE ⟨57-3⟩.

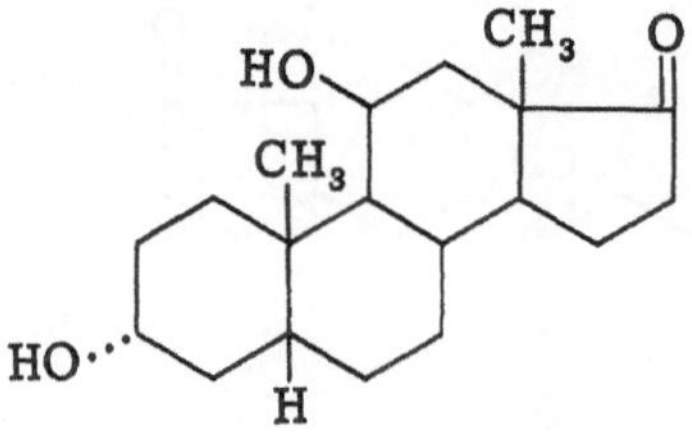

3α,11β-Dihydroxy-5β-androstan-17-one
(11β-Hydroxyetiocholanolone)

3-β-D-GLUCOPYRANOSIDURONIC ACID

2′,3′,4′-Tri-*O*-acetyl methyl ester – $C_{32}H_{46}O_{12}$ (anal. C,H), mp 198-199°, $[\alpha]_D^{20}$ + 35° (chloroform), CC, PC ⟨64-6⟩.

Allied Characterization Data

3-GLUCURONIDE

CC ⟨60-8⟩ ⟨64-6⟩ • PC ⟨64-6⟩ • TLC ⟨66-43⟩.

3-GLUCURONIDE TRI-*O*-ACETYL METHYL ESTER

CC, PC ⟨64-10⟩.

3,11-DISULFATE

CC, PE ⟨57-3⟩.

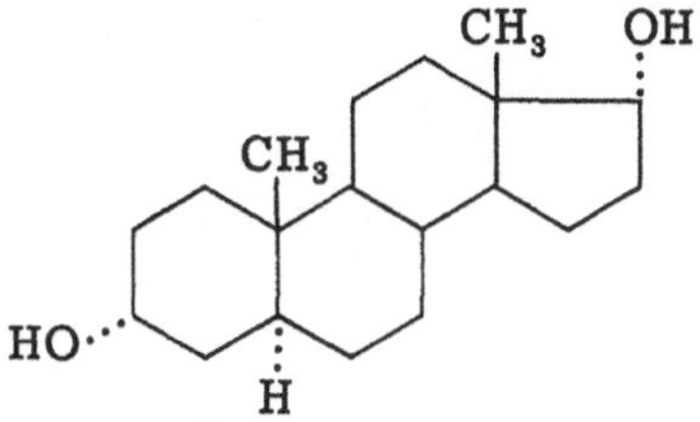

5α-Androstane-3α,17α-diol

SULFATE (Unspecified Location or Number)

Enzymatic synthesis ⟨56-5⟩.

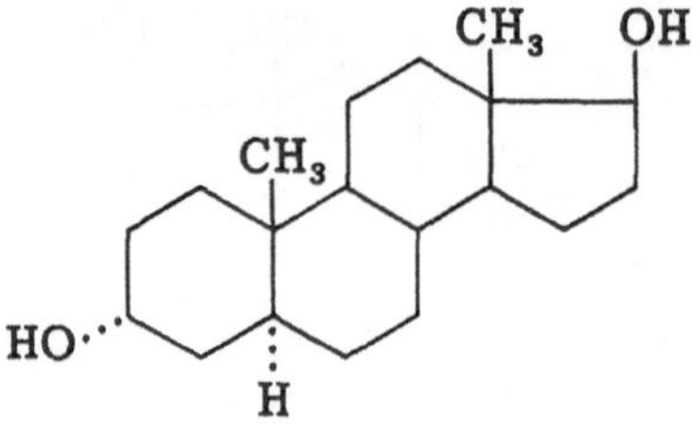

5α-Androstane-3α,17β-diol

GLUCURONIDE (Unspecified Location)

4-^{14}C, TLC ⟨65-24⟩.

SULFATE (Unspecified Location)

TLC ⟨64-3⟩.

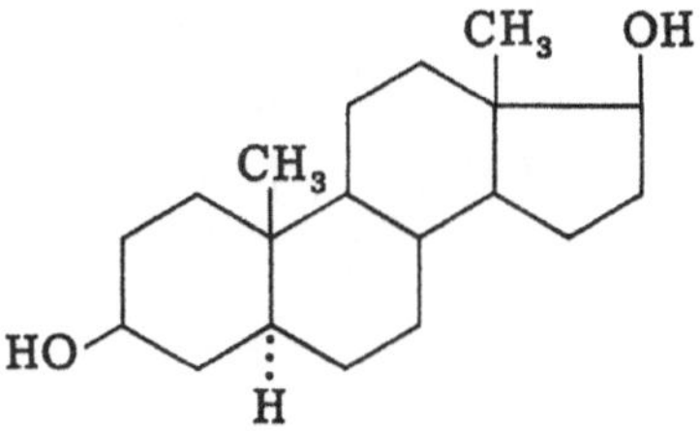

5α-Androstane-3β,17β-diol

GLUCURONIDE (Unspecified Location)

4-^{14}C, TLC ⟨65-24⟩.

3,17-DISULFATE

PC, PE ⟨62-9⟩.

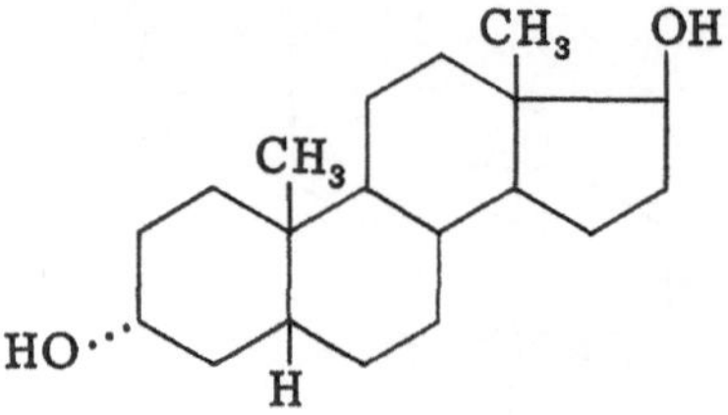

5β-Androstane-3α,17β-diol

17-β-D-GLUCOPYRANOSIDURONIC ACID

7α-^{3}H, 6′-^{14}C, PC ⟨65-23⟩.

Allied Characterization Data

GLUCURONIDE (Unspecified Location)

CC, PC ⟨67-26⟩.

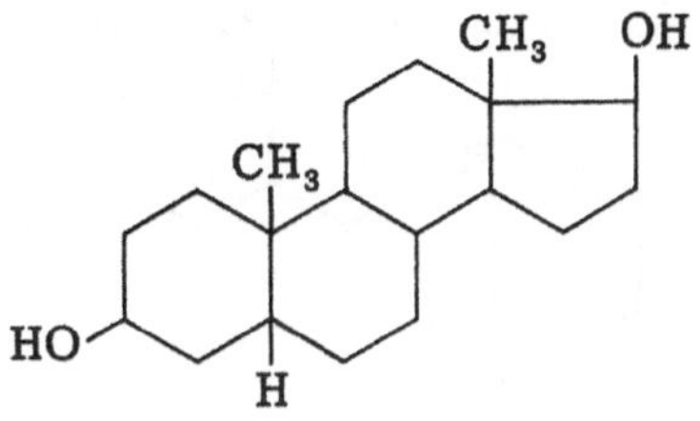

5β-Androstane-3β,17β-diol

GLUCURONIDE (Unspecified Location)

CC, PC ⟨67-26⟩.

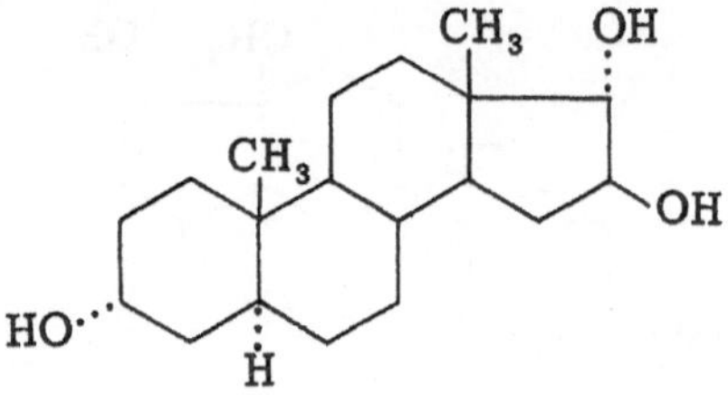

5α-Androstane-3α,16β,17α-triol

SULFATE (Unspecified Location or Number)

Enzymatic synthesis, PC ⟨56-5⟩.

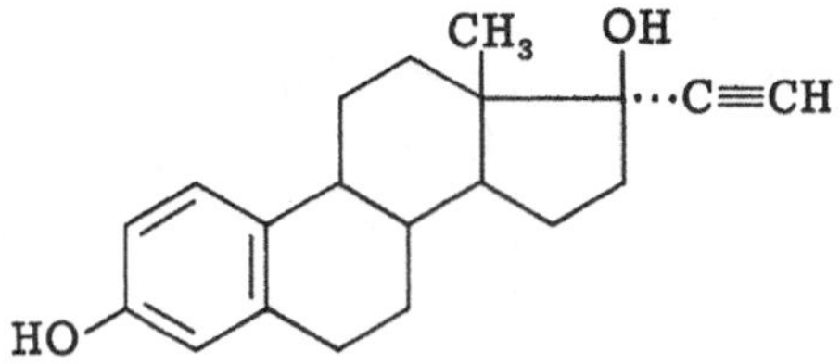

17α-Ethynylestra-1,3,5(10)-triene-3,17β-diol
(Ethynylestradiol)

17-SULFATE-3-METHYL ETHER

Potassium salt – $C_{21}H_{25}KO_5S \cdot \frac{3}{4}H_2O$ (anal. C, H, K, S, H_2O), mp 130° ⟨66-25⟩.

3,17-DISULFATE

Dipotassium salt – $C_{20}H_{22}K_2O_8S_2 \cdot 1\frac{1}{4}H_2O$ (anal. C, H, K, S, H_2O), mp 165-170°, $[\alpha]_D^{25}$ – 2° (dimethylsulfoxide), $\lambda_{max}^{methanol}$ 215, 270 mμ (ε 8700, 790) ⟨66-25⟩.

Disodium salt – $C_{20}H_{22}Na_2O_8S_2 \cdot 1\frac{1}{2}H_2O$ (anal. C, H, Na, S, H_2O), mp 130°, $[\alpha]_D^{25}$ – 6° (methanol), $\lambda_{max}^{methanol}$ 215, 270 mμ (ε 8700, 790) ⟨66-25⟩.

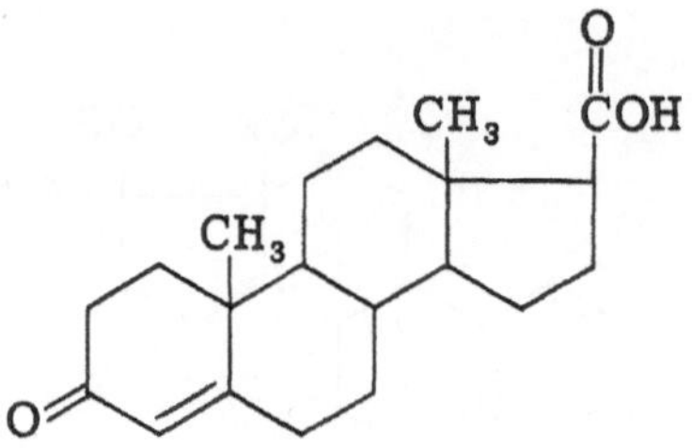

17β-Carboxyandrost-4-en-3-one
(3-Keto-4-etienic acid)

20-β-D-GLUCOPYRANOSIDURONIC ACID ($C_{20} \rightarrow C_{1'}$ ESTER)

Acid – PC ⟨64-24⟩.

Methyl ester – $C_{27}H_{38}O_9$ (anal. C, H), $[\alpha]_D^{26}$ + 113° (methanol), λ_{max} 241-242° mμ (ε 16,050) ⟨64-24⟩.

Methyl ester, 3-(2,4,dinitrophenylhydrazone) – $C_{33}H_{42}N_4O_{12}$ (anal. C, H, N), mp 222-224°, λ_{max} 392 mμ (ε 28,000) ⟨64-24⟩.

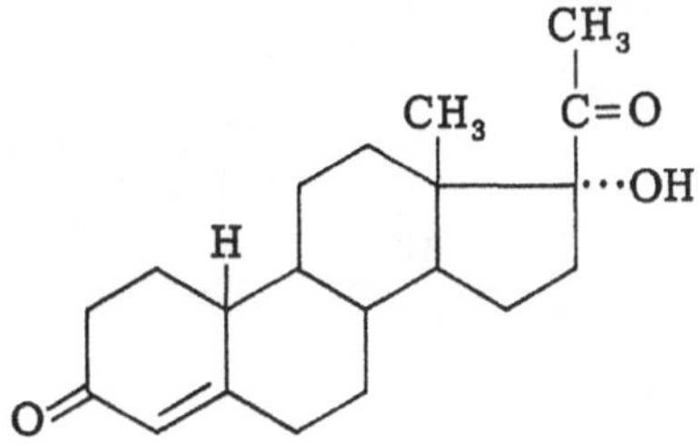

17α-Hydroxy-19-norpregn-4-ene-3,20-dione

17-SULFATE

Sodium salt — mp 152-155°, $[\alpha]_D^{22}$ + 63.7° (methanol), λ_{max} 239 mμ (ε 15,920) ⟨63-6⟩ ⟨64-23⟩.

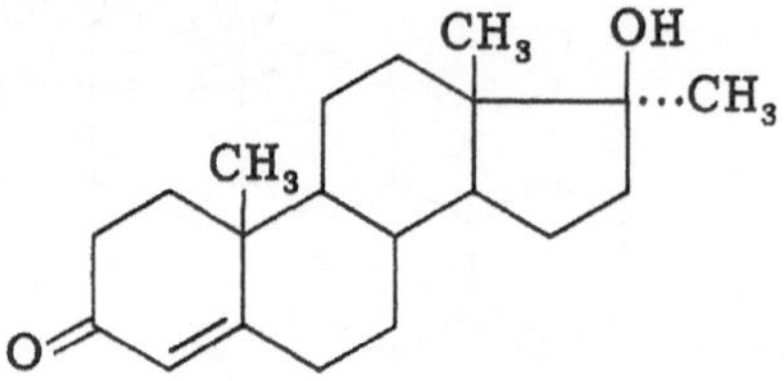

17β-Hydroxy-17α-methylandrost-4-en-3-one
(Methyltestosterone)

β-D-GLUCOPYRANOSIDURONIC ACID (Unspecified Location)

Enzymatic synthesis ⟨57-10⟩.

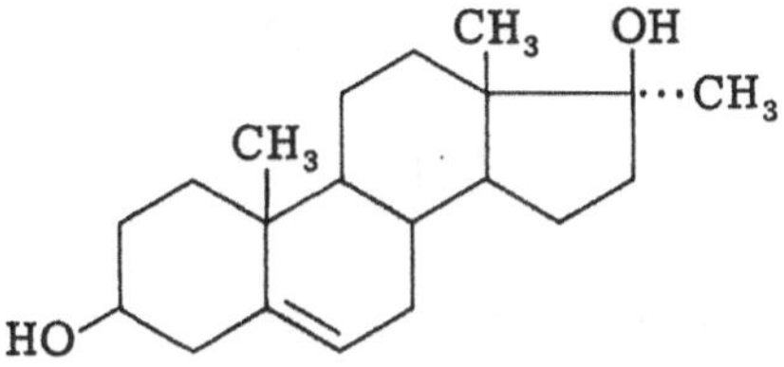

17α-Methylandrost-5-ene-3β,17β-diol
(Methylandrostenediol)

3-SULFATE

Sodium salt – mp 152-154°, mp 173-175°, PC, PE ⟨57-2⟩.

3.17-DISULFATE

Disodium salt – mp 141-145°, $[\alpha]_D^{20}$ – 36.3° (methanol) ⟨57-5⟩ • mp 140-142° ⟨57-2⟩.

Allied Characterization Data

3-SULFATE

Enzymatic synthesis, PC ⟨56-5⟩.

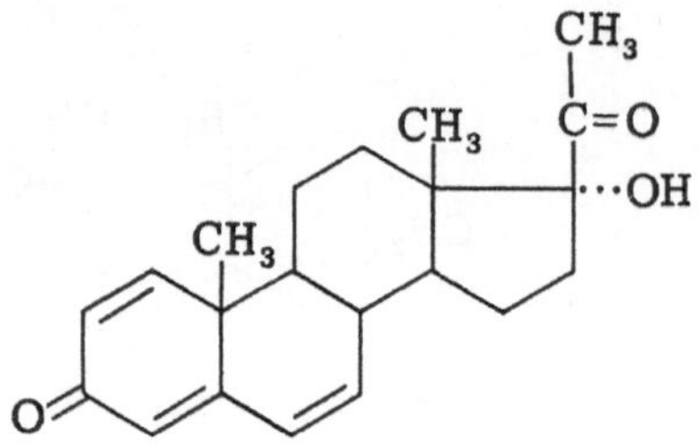

17α-Hydroxypregna-1,4,6-triene-3,20-dione

17-SULFATE

Sodium salt — mp 166° ⟨63-6⟩.

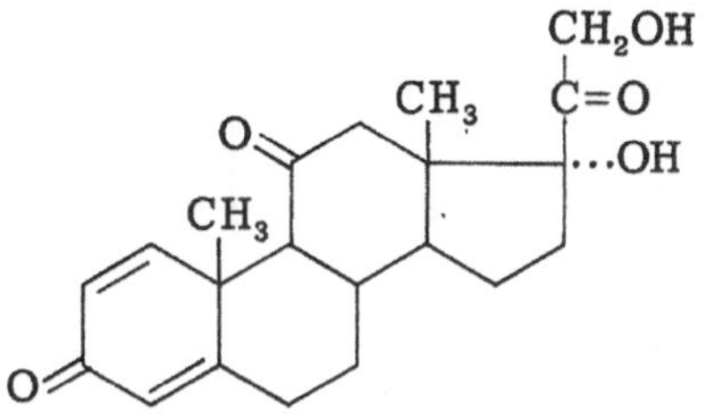

17α,21-Dihydroxypregna-1,4-diene-3,11,20-trione
(Prednisone)

21-β-D-GLUCOPYRANOSIDURONIC ACID

Amide — $C_{27}H_{35}NO_{10}$ (anal. C, H, N), mp 257-258°, $[\alpha]_D^{24}$ + 97° (dioxane-ethanol 1:1), $\lambda_{max}^{ethanol}$ 242 mμ (ε 0.941 × 10^4), IR ⟨64-14⟩.

2′,3′,4′-Tri-*O*-acetyl methyl ester — $C_{34}H_{42}O_{14} \cdot H_2O$ (anal. C, H), mp 135-137°, $[\alpha]_D^{19}$ + 95° (ethanol), $\lambda_{max}^{ethanol}$ 238 mμ (ε 1.906 × 10^4), IR ⟨64-14⟩ • mp 135-136° ⟨62-11⟩.

21-SULFATE

Pyridine salt — mp 160°, λ_{max} 240 mμ (ε 15,730) ⟨60-9⟩.

Sodium salt — mp 158-161°, $[\alpha]_D$ + 138° (water), pH of 0.5%, aqueous solution 7.2 ⟨60-9⟩.

Triethylammonium salt — $C_{27}H_{41}NO_8S$, mp 201-202°, $[\alpha]_D$ + 131° (water), $\lambda_{max}^{methanol}$ 238 mμ (ε 15,200), IR ⟨65-19⟩ • mp 201-202° ⟨60-9⟩.

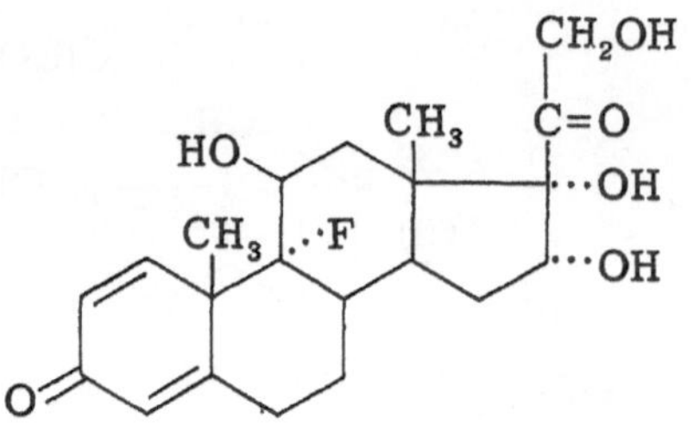

$C_{21}H_{27}FO_6$

9α-Fluoro-11β,16α,17α,21-tetrahydroxypregna-1,4-diene-3,20-dione
(Triamcinolone)

21-SULFATE-16,17-ACETONIDE

N,N′-Dibenzylethylenediamine salt — mp 230-232°, $[\alpha]_D^{26}$ + 90° (ethanol) ⟨64-19⟩.

Sodium salt — mp 300-305°, $[\alpha]_D^{24}$ + 95° (water) ⟨64-19⟩.

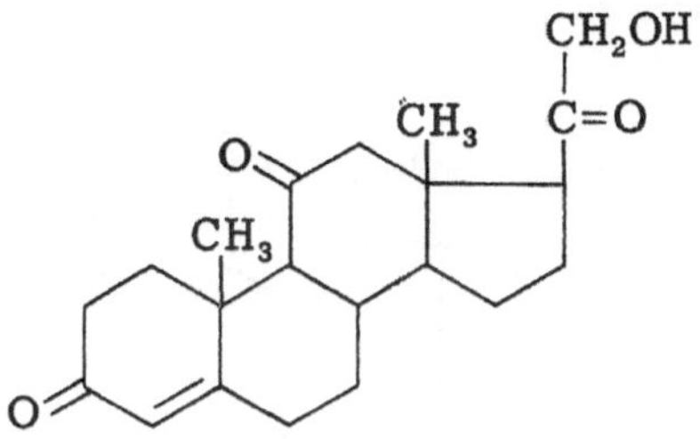

21-Hydroxypregn-4-ene-3,11,20-trione
(Dehydrocorticosterone)

21-SULFATE

Pyridine salt – $C_{26}H_{33}NO_7S$ (anal. S), mp 175-176°, $[\alpha]_D^{22}$ + 142° (methanol), IR, PC, PE ⟨62-10⟩ • PC ⟨60-15⟩.

Allied Characterization Data

21-SULFATE

PC, PE ⟨64-1⟩.

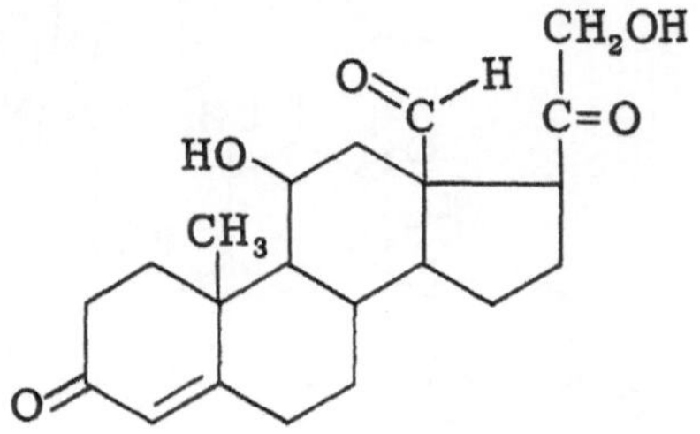

11β,21-Dihydroxy-3,20-dioxopregn-4-en-18-al
(Aldosterone)

GLUCURONIDE

CC ⟨64-17⟩ ⟨65-21⟩ • PC ⟨64-15⟩ ⟨64-17⟩ ⟨65-21⟩ • PE ⟨64-17⟩ ⟨65-21⟩.

GLUCURONIDE-21-ACETATE

PC ⟨64-17⟩.

17α,21-Dihydroxypregn-4-ene-3,11,20-trione
(Cortisone)

21-β-D-GLUCOPYRANOSIDURONIC ACID

Acid – ⟨64-27⟩ • mp 172-173°, $[\alpha]_D$ + 93.5° (methanol), $\lambda_{max}^{methanol}$ 238 mμ (ε 16,400) ⟨65-18⟩.

Amide – $C_{27}H_{37}NO_{10}\cdot H_2O$ (anal. C, H, N), mp 244-245°, $[\alpha]_D^{24}$ + 99°, $\lambda_{max}^{ethanol}$ 241 mμ (ε 0.536 × 10^4), IR ⟨64-14⟩.

Methyl ester – mp 142-143°, mp 191-193°, $[\alpha]_D$ + 93° (methanol) ⟨65-18⟩.

2′,3′,4′-Tri-*O*-acetyl methyl ester – $C_{34}H_{44}O_{14}$ (anal. C, H), mp 105-107°, $[\alpha]_D^{24}$ + 120° (chloroform) ⟨59-7⟩ • (anal. C, H), mp 127-129°, $[\alpha]_D^{18}$ + 103.1° (chloroform), $\lambda_{max}^{methanol}$ 239 mμ (log ε 4.1) ⟨64-27⟩ • (anal. C, H), mp 193°, $[\alpha]_D^{19}$ + 95° (ethanol), $\lambda_{max}^{ethanol}$ 237 mμ (ε 1.699 × 10^4), IR ⟨64-14⟩ • mp 197-198°, $[\alpha]_D$ + 87° (chloroform), $\lambda_{max}^{methanol}$ 238 mμ (ε 16,000) ⟨65-18⟩ • mp 193° ⟨62-11⟩.

21-SULFATE

Ammonium salt – $C_{21}H_{31}NO_8S$ (anal. C, H, N, S), mp > 275°, $[\alpha]_D^{25}$ + 153° (methanol), $\lambda_{max}^{methanol}$ 238 mμ (ε 15,600), NMR ⟨68-2⟩.

Pyridine salt – $C_{26}H_{33}NO_8S$ (anal. S), mp 195-198°, $[\alpha]_D^{22}$ + 136° (methanol), IR, PC, PE ⟨62-10⟩ • PC ⟨60-15⟩.

Sodium salt – PC ⟨60-15⟩.

Triethylammonium salt – $C_{27}H_{43}NO_8S$ (anal. C, H, N, S), mp 226-227°, $[\alpha]_D^{25}$ + 127° (methanol), $\lambda_{max}^{methanol}$ 238 mμ (ε 15,400), NMR ⟨68-2⟩.

Allied Characterization Data

21-SULFATE

PC, PE ⟨62-9⟩ ⟨64-1⟩.

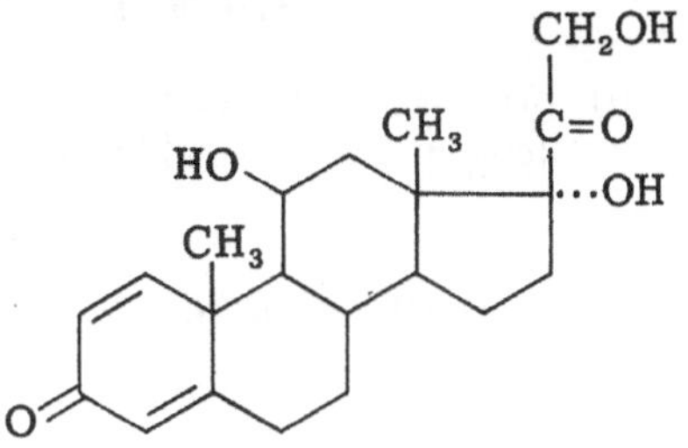

$11\beta,17\alpha,21$-Trihydroxypregna-1,4-diene-3,20-dione
(Prednisolone)

21-2′-AMINO-2′-DEOXY-β-D-GLUCOPYRANOSIDE

2′-Acetyl — $C_{29}H_{41}NO_{10}$ (anal. C, H, N), mp 183-184°, $[\alpha]_D^{25}$ + 66° (methanol), $\lambda_{max}^{methanol}$ 243 mμ (log ε 4.19) ⟨64-9⟩.

2′,3′,4′,6′-Tetraacetyl — $C_{35}H_{47}NO_{13}$ (anal. C, H), mp 246-248°, $[\alpha]_D^{25}$ + 36° (chloroform), $\lambda_{max}^{methanol}$ 243 mμ (log ε 4.17) ⟨64-9⟩.

21-β-D-GLUCOPYRANOSIDURONIC ACID

Amide — $C_{27}H_{37}NO_{10}\cdot H_2O$ (anal. C, H, N), mp 253-254°, $[\alpha]_D^{24}$ + 62° (dioxane-ethanol 1:1), $\lambda_{max}^{ethanol}$ 248 mμ (ε 0.672 × 10^4), IR ⟨64-14⟩.

Sodium salt — λ_{max} 246 mμ ⟨63-15⟩.

2′,3′,4′-Tri-*O*-acetyl methyl ester — $C_{34}H_{44}O_{14}\cdot H_2O$ (anal. C, H), mp 118-122°, $[\alpha]_D^{19}$ + 59° (ethanol), $\lambda_{max}^{ethanol}$ 244 mμ (ε 1.237 × 10^4), IR ⟨64-14⟩ • λ_{max} 243 mμ ⟨63-15⟩ • mp 130-132° ⟨62-11⟩.

21-SULFATE

Pyridine salt — mp 160°, λ_{max} 240 mμ (ε 15,730) ⟨64-7⟩.

Sodium salt — mp 179°, $[\alpha]_D^{21}$ + 101.8° (water), λ_{max} 242 mμ

(ϵ 14,740), pH of 0.5% aqueous solution 7.35 ⟨59-2⟩ ⟨64-7⟩ • $[\alpha]_D$ + 110°, λ_{max} 239 mμ ($E^{1\%}_{1 cm.}$ 312) ⟨65-26⟩.

Triethylammonium salt – mp 201-202° ⟨64-7⟩.

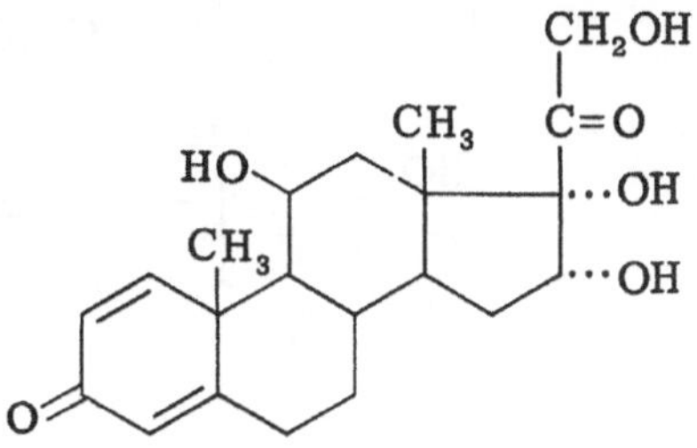

11β,16α,17α,21-Tetrahydroxypregna-1,4-diene-3,20-dione
(16α-Hydroxyprednisolone)

21-SULFATE

Sodium salt – $C_{21}H_{27}NaO_9S$, mp 200°, $[\alpha]_D$ + 66° (water), λ_{max}238 mμ (ε 13,900), pH 6.5 (0.5% aqueous solution) ⟨60-9⟩.

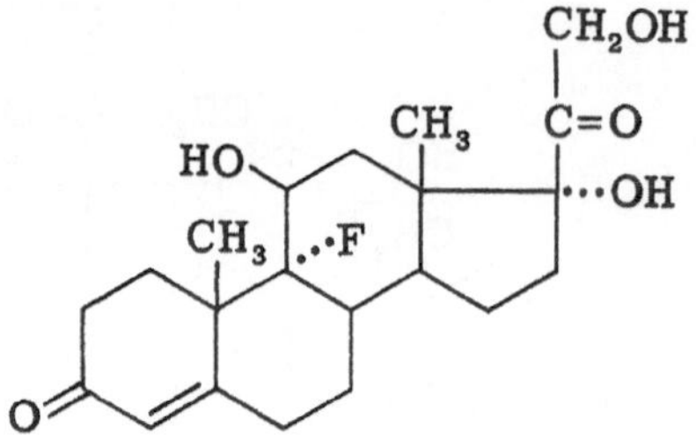

9α-Fluoro-11β,17α,21-trihydroxypregn-4-ene-3,20-dione
(9α-Fluorohydrocortisone, 9α-Fluorocortisol)

21-SULFATE

Sodium salt – mp 190°, $[\alpha]_D$ + 118.2°, λ_{max} 239 mμ (ϵ 15,820), pH 7.4 (0.5% aqueous solution) ⟨60-9⟩ ⟨64-7⟩.

Triethylammonium salt – mp 202-203°, $[\alpha]_D$ + 104.3° (water), λ_{max} 239 mμ (ϵ 16,750) ⟨60-9⟩ • mp 195-196°, $[\alpha]_D$ + 104° (water), λ_{max} 239 mμ (ϵ 16,820) ⟨64-7⟩.

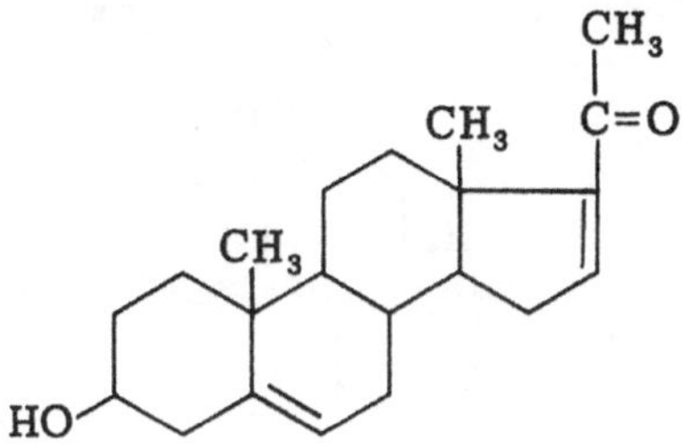

3β-Hydroxypregna-5,16-dien-20-one

3-SULFATE

Enzymatic synthesis, PC ⟨56-5⟩.

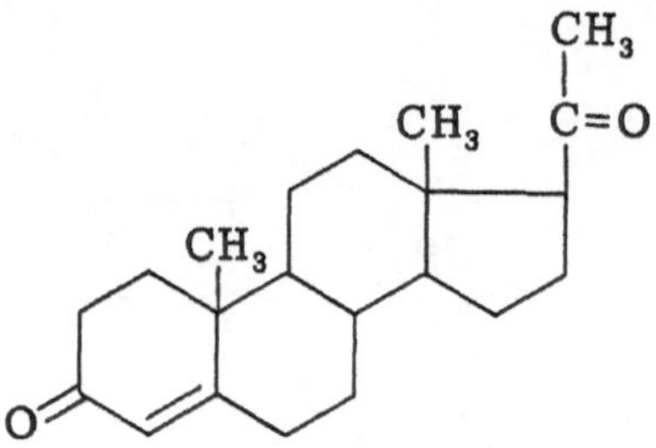

Pregn-4-ene-3,20-dione
(Progesterone)

3,5-DIENOL-3-β-D-GLUCOPYRANOSIDURONIC ACID

γ-Lactone – $C_{27}H_{36}O_7$ (anal. C, H), mp 216-224° ⟨59-7⟩.

2′,3′,4′-Tri-*O*-acetyl methyl ester – $C_{34}H_{46}O_{11}$ (anal. C, H), mp 191-194°, $[\alpha]_D^{27}$ ± 0° (chloroform) ⟨59-7⟩ • IR ⟨59-8⟩ • mp 187-188° ⟨57-11⟩.

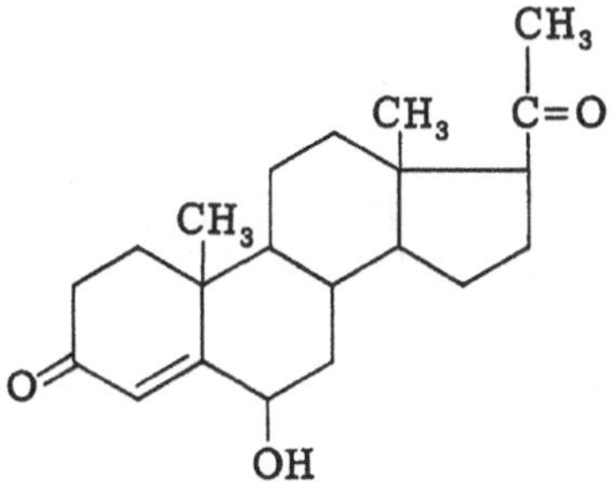

6β-Hydroxypregn-4-ene-3,20-dione
(6β-Hydroxyprogesterone)

6-SULFATE

Triethylammonium salt – $C_{27}H_{45}NO_6S$ (anal. C, H, N, S), mp 180-181°, $[\alpha]_D^{25}$ + 59° (chloroform), $\lambda_{max}^{methanol}$ 236 mμ (ε 13,300), NMR ⟨68-2⟩.

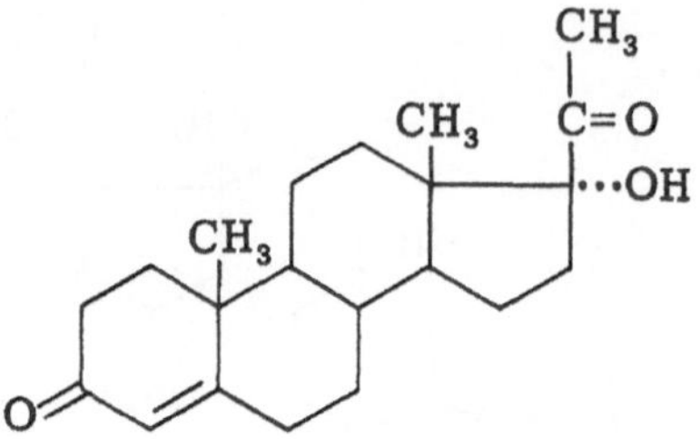

17α-Hydroxypregn-4-ene-3,20-dione
(17α-Hydroxyprogesterone)

17-SULFATE

Pyridine salt – mp 188-193°, $[\alpha]_D$ + 96° (methanol), λ_{max} 240 mμ (ϵ 16,000) ⟨63-6⟩.

Sodium salt – mp 155-160°, $[\alpha]_D$ + 121° (water), λ_{max} 240 mμ (ϵ 16,770) ⟨63-6⟩.

Triethylammonium salt – $C_{27}H_{45}NO_6S$ (anal. C, H, N, S), mp 209-212°, $[\alpha]_D^{25}$ + 104° (chloroform), $\lambda_{max}^{methanol}$ 241 mμ (ϵ 17,600), NMR ⟨68-2⟩.

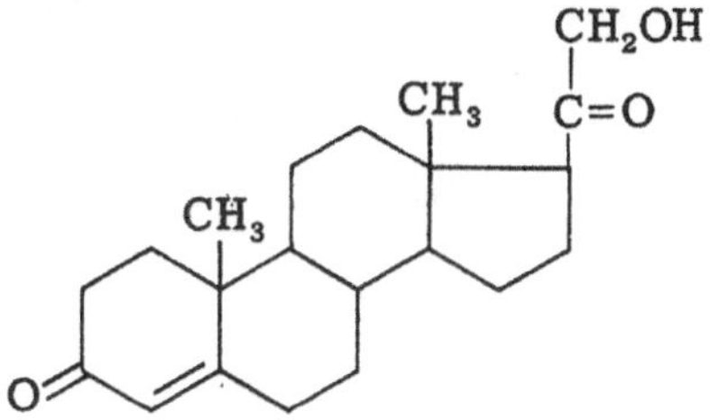

21-Hydroxypregn-4-ene-3,20-dione
(Deoxycorticosterone, Cortexone)

21-β-D-GLUCOPYRANOSIDURONIC ACID

Sodium salt — $\lambda_{max}^{methanol}$ 240 mμ ⟨59-3⟩.

2′,3′,4′-Tri-*O*-acetyl methyl ester — $C_{34}H_{46}O_{12}$ (anal. C, H), mp 205-207.5°, $[\alpha]_D^{20}$ + 68° (chloroform), $\lambda_{max}^{alcohol}$ 240 mμ (log ε 4.2) ⟨58-10⟩ • $C_{34}H_{46}O_{12}$ (anal. C, H, O), mp 199.5-201.5°, λ_{max} 240 mμ (ε 17,200) ⟨59-3⟩.

21-SULFATE

Pyridine salt — $C_{26}H_{35}NO_6S$ (anal. S), mp 185-188°, $[\alpha]_D^{22}$ + 107° (95% ethanol), IR, PC, PE ⟨62-10⟩ • PC ⟨60-15⟩ ⟨61-13⟩ • PE ⟨61-13⟩.

Sodium salt — mp 190-195° ⟨57-2⟩ • PC ⟨59-3⟩ ⟨60-15⟩ ⟨61-13⟩ • PE ⟨57-2⟩ ⟨61-13⟩.

Triethylammonium salt — $C_{27}H_{45}NO_6S$ (anal. C, H, N, S), mp 141-142°, $[\alpha]_D^{25}$ + 120° (chloroform), $\lambda_{max}^{methanol}$ 238 mμ (ε 19,400), NMR ⟨68-2⟩.

Allied Characterization Data

21-SULFATE

CC ⟨65-3⟩ • IR ⟨65-14⟩ ⟨67-18⟩ • PC ⟨56-5⟩ ⟨62-9⟩ ⟨64-1⟩ ⟨65-14⟩ ⟨66-28⟩ ⟨67-18⟩ • PE ⟨62-9⟩ ⟨64-1⟩ • TLC ⟨65-14⟩ ⟨66-28⟩ ⟨67-18⟩.

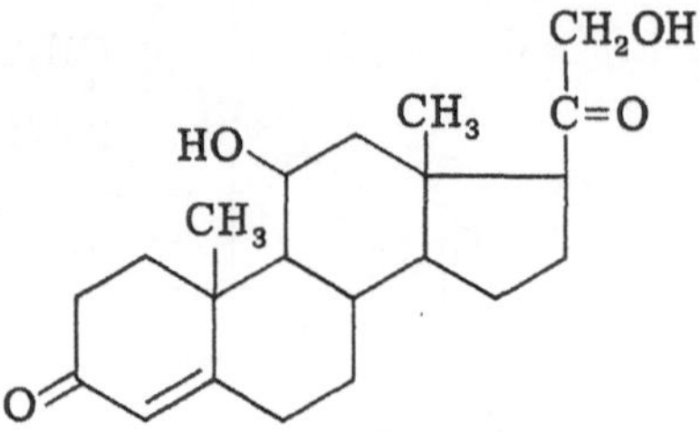

11β,21-Dihydroxypregn-4-ene-3,20-dione
(Corticosterone)

21-SULFATE

Potassium salt – $C_{21}H_{29}KO_7S$ (anal. S), mp 178-180°, PC, PE ⟨62-10⟩.

Pyridine salt – PC ⟨60-15⟩ ⟨61-13⟩ • PE ⟨61-13⟩.

Triethylammonium – $C_{27}H_{45}NO_7S$ (anal. C, H, N, S), mp 150-151°, $[\alpha]_D^{25}$ + 147° (chloroform), $\lambda_{max}^{methanol}$ 242 mμ (ε 16,100), NMR ⟨68-3⟩.

Allied Characterization Data

21-SULFATE

CC ⟨60-14⟩ ⟨65-3⟩ ⟨65-20⟩ • IR ⟨65-14⟩ ⟨66-26⟩ ⟨67-18⟩ • PC ⟨65-14⟩ ⟨65-20⟩ ⟨66-26⟩ ⟨66-28⟩ ⟨67-18⟩ • TLC ⟨65-14⟩ ⟨66-26⟩ ⟨66-28⟩ ⟨67-18⟩.

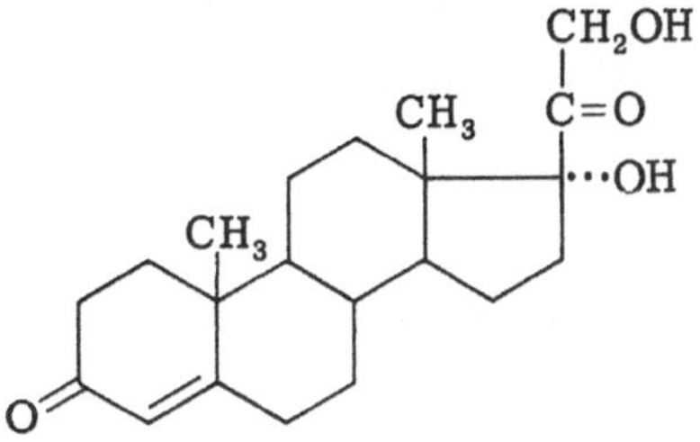

17α,21-Dihydroxypregn-4-ene-3,20-dione
(Compound S)

21-β-D-GLUCOPYRANOSIDURONIC ACID

Amide – $C_{27}H_{39}NO_9 \cdot H_2O$ (anal. C, H, N), mp 242.5-243°, $[\alpha]_D^{24}$ + 52° (ethanol-dioxane 1:1), $\lambda_{max}^{ethanol}$ 240 mμ (ε 1.017 × 10^4), IR ⟨64-14⟩.

2′,3′,4′-Tri-*O*-acetyl methyl ester – $C_{34}H_{46}O_{13}$ (anal. C, H), mp 134-136°, $[\alpha]_D^{19}$ + 48° (ethanol), $\lambda_{max}^{ethanol}$ 242 mμ, (ε 1.040 × 10^4), IR ⟨64-14⟩ • mp 135.5-137° ⟨62-11⟩.

21-SULFATE

Pyridine salt – $C_{26}H_{35}NO_7S$ (anal. S), mp 176-178°, $[\alpha]_D^{22}$ + 89° (methanol), IR, PC, PE ⟨62-10⟩ • PC ⟨60-15⟩.

Triethylammonium salt – $C_{27}H_{45}NO_7S$ (anal. C, H, N, S), mp 224-225°, $[\alpha]_D^{25}$ + 88° (chloroform), $\lambda_{max}^{methanol}$ 241 mμ (ε 16,600), NMR ⟨68-2⟩.

17,21-DISULFATE

Di-triethylammonium salt – $C_{33}H_{60}N_2O_{10}S_2$ (anal. C, H, N, S), mp 201-202°, $[\alpha]_D^{25}$ + 90° (chloroform), $\lambda_{max}^{methanol}$ 242 mμ (ε 16,600), NMR ⟨68-2⟩.

Allied Characterization Data

21-SULFATE

PC ⟨63-11⟩⟨64-1⟩⟨65-14⟩⟨66-28⟩ • PE ⟨63-11⟩⟨64-1⟩ • TLC ⟨65-14⟩⟨66-28⟩.

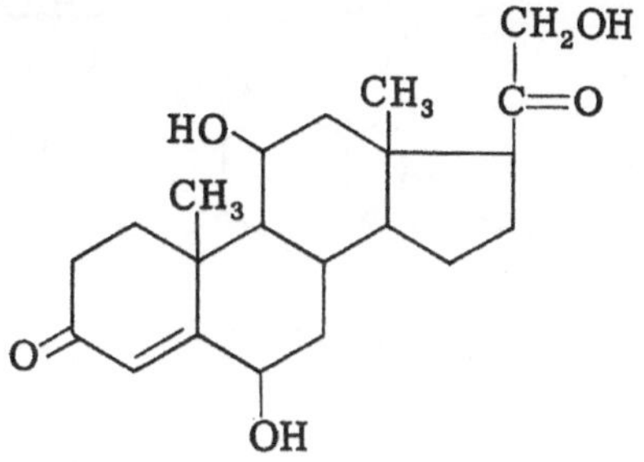

6β,11β,21-Trihydroxypregn-4-ene-3,20-dione
(6β-Hydroxycorticosterone)

21-SULFATE

CC ⟨64-16⟩.

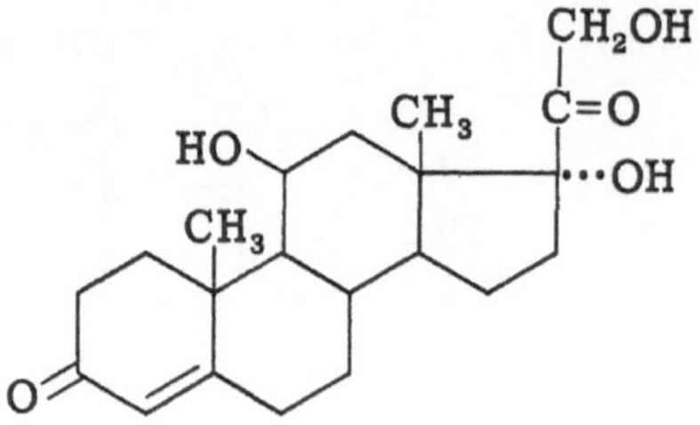

11β,17α,21-Trihydroxypregn-4-ene-3,20-dione
(Hydrocortisone, Cortisol)

21-β-D-GLUCOPYRANOSIDURONIC ACID

Acid – mp 159-160° ⟨65-18⟩

Amide – $C_{27}H_{39}NO_{10}\cdot H_2O$ (anal. C, H, N), mp 259-260°, $[\alpha]_D$ + 67° (dioxane-ethanol 1:1), $\lambda_{max}^{ethanol}$ 244 mμ (ε 1.204 × 10^4), IR ⟨64-14⟩.

Methyl ester – mp 144-147°, $[\alpha]_D$ + 77° (methanol) ⟨65-18⟩.

2′,3′,4′-Tri-*O*-acetyl methyl ester – $C_{34}H_{46}O_{14}$ (anal. C, H), mp 118-120°, $[\alpha]_D^{19}$ + 69° (ethanol), $\lambda_{max}^{ethanol}$ 242 mμ (ε 1.779 × 10^4), IR ⟨64-14⟩ • mp 122-124°, $[\alpha]_D$ + 72° (chloroform), $\lambda_{max}^{methanol}$ 242 mμ (ε 16,400) ⟨65-18⟩ • mp 130-132° ⟨62-11⟩.

11-SULFATE-21-ACETATE

Pyridine salt – mp 150°, $[\alpha]_D$ + 135° (water) ⟨62-7⟩.

Sodium salt – mp 160°, $[\alpha]_D$ + 160° (water), λ_{max} 240 mμ (ε 16,000) ⟨62-7⟩.

21-SULFATE

Choline salt – mp 225-227°, $[\alpha]_D$ + 119.3° (water) ⟨64-7⟩.

Potassium salt – mp 179°, $[\alpha]_D$ + 131.8° (water) ⟨64-7⟩.

Pyridine salt – PC ⟨60-15⟩⟨61-13⟩ • PE ⟨61-13⟩.

Sodium salt – $C_{21}H_{29}NaO_8S$ (anal. C, H, S), mp 146-148°, CC, PE ⟨64-12⟩ • mp 185°, $[\alpha]_D$ + 130.5° (water), pH 7.0 (0.5% aqueous solution) ⟨58-5⟩ • mp 178°, $[\alpha]_D$ + 128° (water) ⟨64-7⟩ • PC, PE ⟨57-2⟩.

Tetramethylammonium salt – mp 230°, $[\alpha]_D$ + 126.3° (water) ⟨64-7⟩.

Triethylammonium salt – $C_{27}H_{45}NO_8S$ (anal. C, H, N, S), mp 189-191°, $[\alpha]_D^{25}$ + 110° (chloroform), $\lambda_{max}^{methanol}$ 241 mμ (ϵ 16,850), NMR ⟨68-2⟩ • mp 191-193°, $[\alpha]_D$ + 117.8° (water) ⟨64-7⟩.

Allied Characterization Data

21-SULFATE

PC ⟨64-1⟩ • PE ⟨64-1⟩⟨64-11⟩.

DISULFATE (Unspecified Location)

PC, PE ⟨57-2⟩.

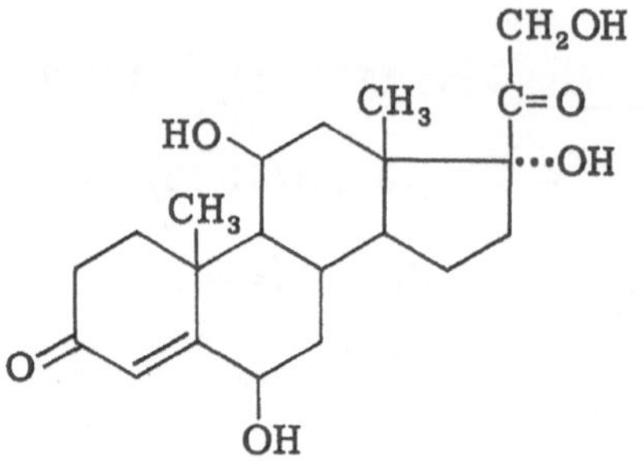

6β,11β,17α,21-Tetrahydroxypregn-4-ene-3,20-dione
(6β-Hydroxycortisol, 6β-Hydroxyhydrocortisone)

6-SULFATE-21-ACETATE

Triethylammonium salt – $C_{29}H_{47}NO_{10}S$ (anal. N, S), mp 190-192°, $[\alpha]_D^{25}$ + 60° (chloroform), $\lambda_{max}^{methanol}$ 235 mμ (ε 15,400) ⟨68-3⟩.

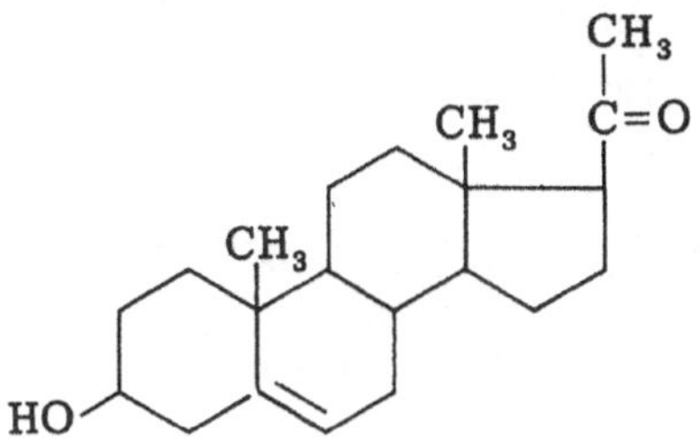

3β-Hydroxypregn-5-en-20-one
(Pregnenolone)

3-β-D-GLUCOPYRANOSIDURONIC ACID

Acid – $C_{27}H_{40}O_8 \cdot H_2O$ (anal. C, H), mp 245-248°, $[\alpha]_D^{21}$ – 40° (methanol) ⟨59-7⟩.

2′,3′,4′-Tri-*O*-acetyl methyl ester – $C_{34}H_{48}O_{11}$ (anal. C, H), mp 183-184°, $[\alpha]_D^{23}$ + 5° (chloroform) ⟨59-7⟩ • IR ⟨59-8⟩.

Allied Characterization Data

3-GLUCURONIDE

PC ⟨66-12⟩.

* * * * *

3-SULFATE

Ammonium salt – $C_{21}H_{35}NO_5S \cdot H_2O$ (anal. C, H, N, S, H_2O), mp 206-207°, $[\alpha]_D^{25}$ + 19° (chloroform) ⟨66-25⟩ • mp 195-198°, mp 198-201°, ^{35}S and 7α-^3H, CC, IR ⟨66-3⟩ • mp 198-201°, CC, IR ⟨63-3⟩ • no physical constants ⟨65-6⟩.

Methyl ester – $C_{22}H_{34}O_5S$ (anal. C, H, S), mp 118°, $[\alpha]_D^{23}$ + 22 (chloroform) ⟨60-18⟩.

Potassium salt – $C_{21}H_{31}KO_5S \cdot 1\frac{1}{4}H_2O$ (anal. C, H, S, K, H_2O), mp 210-212°, $[\alpha]_D^{25}$ + 24° (methanol) ⟨66-25⟩, mp 226-228° ⟨57-9⟩.

Pyridine salt – $C_{26}H_{37}NO_5S$ (anal. N, S), mp 185-186°, $[\alpha]_D$ + 16° (chloroform) ⟨57-9⟩ • (anal. C, H, N), mp 173°, $[\alpha]_D^{23}$ + 23° (chloroform) ⟨60-18⟩.

Sodium salt – PC ⟨55-2⟩.

Allied Characterization Data

3-SULFATE

CC ⟨64-2⟩⟨66-5⟩⟨67-32⟩ • IR ⟨67-18⟩ • PC ⟨56-5⟩ ⟨67-5⟩⟨67-18⟩ • PE ⟨64-11⟩ • TLC ⟨66-21⟩⟨67-18⟩ ⟨67-24⟩⟨67-32⟩.

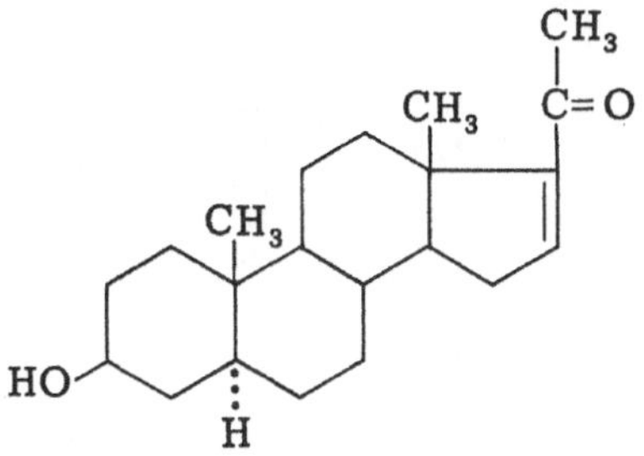

3β-Hydroxy-5α-pregn-16-en-20-one

3-SULFATE

Potassium salt – $C_{21}H_{31}KO_5S \cdot H_2O$ (anal. C, H, K), mp 244-245° (ethanol), mp 216-217° (water) ⟨48-1⟩.

Toluidine salt – $C_{28}H_{41}NO_5S$ (anal. C, H, N, S), mp 195-197°, $[\alpha]_D^{19}$ + 31.7° (chloroform) ⟨48-1⟩.

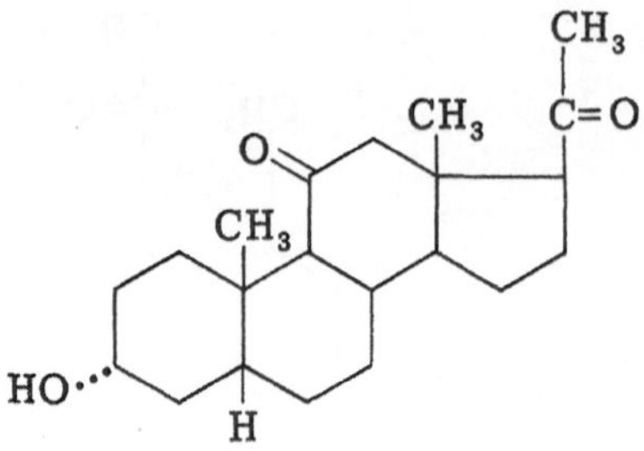

3 α-Hydroxy-5β-pregnane-11,20-dione

3-SULFATE

Sodium salt – $C_{21}H_{31}NaO_6S \cdot 3H_2O$ (anal. C, H), mp 160-162°, $[\alpha]_D$ + 107.5° (water), IR ⟨65-4⟩.

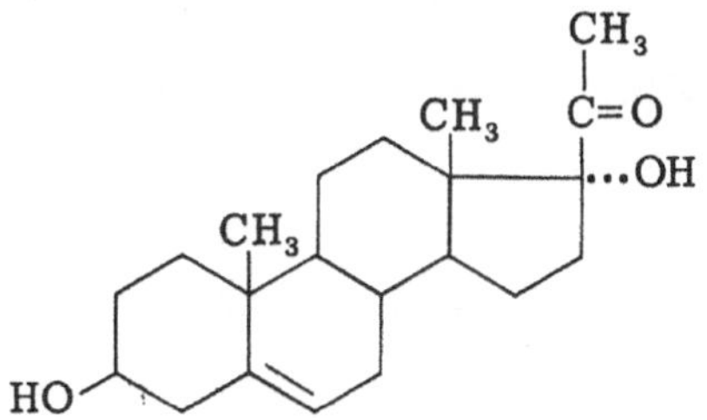

3β,17α-Dihydroxypregn-5-en-20-one
(17α-Hydroxypregnenolone)

3-SULFATE

Ammonium salt – mp 195-198°, CC, IR, PC, 7α-^{3}H, ^{35}S ⟨64-2⟩ • mp 192-194°, CC, IR ⟨66-4⟩.

Pyridine salt – mp 192-195°, CC, IR ⟨64-2⟩.

Triethylammonium salt – $C_{27}H_{47}NO_6S$ (anal. C, H, N, S), mp 235-238°, $[\alpha]_D^{25}$ – 17° (methanol) ⟨68-2⟩.

3-SULFATE-17-ACETATE

Ammonium salt – mp 261-263° ⟨66-1⟩ • no physical constants given ⟨65-6⟩.

Pyridine salt – no physical constants given ⟨65-6⟩.

3,17-DISULFATE

Diammonium salt – $C_{21}H_{38}N_2O_9S_2$ (anal. C, H, N, S), mp 202-204°, $[\alpha]_D^{25}$ + 8° (methanol), NMR ⟨68-2⟩.

Allied Characterization Data

3-SULFATE

CC ⟨66-5⟩ • TLC ⟨66-21⟩⟨67-27⟩.

3-SULFATE-17-ACETATE

CC, TLC ⟨67-32⟩.

$$\text{structure: } 3\beta,21\text{-dihydroxypregn-5-en-20-one } (CH_3,\ CH_3,\ C{=}O,\ CH_2OH,\ HO)$$

3β,21-Dihydroxypregn-5-en-20-one

3,21-DISULFATE

Diammonium salt – ⟨62-9⟩.

Disodium salt – PC, PE ⟨66-34⟩.

Allied Characterization Data

3,21-DISULFATE

CC ⟨62-9⟩⟨63-11⟩ • PC, PE ⟨62-9⟩⟨63-11⟩.

SULFATE (Not Specified Mono or Di)

Enzymatic synthesis, PC ⟨56-5⟩.

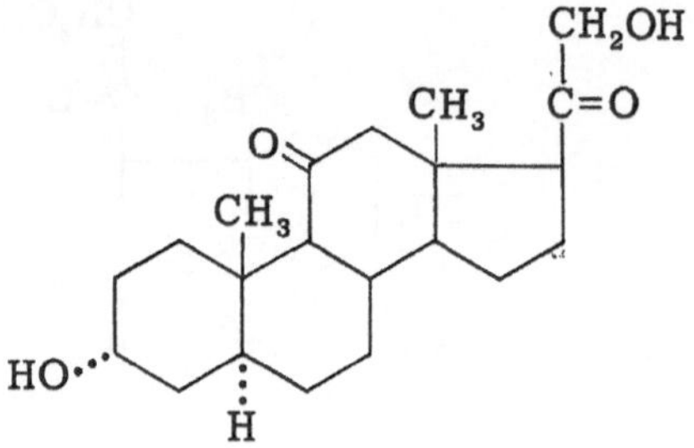

$C_{21}H_{32}O_4$

3α,21-Dihydroxy-5α-pregnane-11,20-dione

3-GLUCURONIDE

PC ⟨63-12⟩.

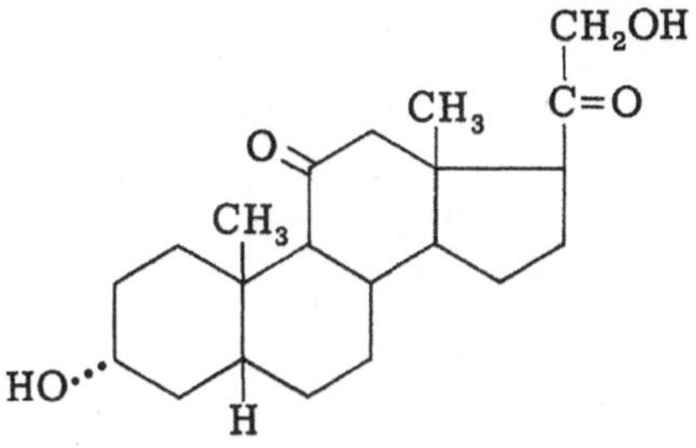

$C_{21}H_{32}O_4$

3α,21-Dihydroxy-5β-pregnane-11,20-dione

3-GLUCURONIDE

PC ⟨63-12⟩.

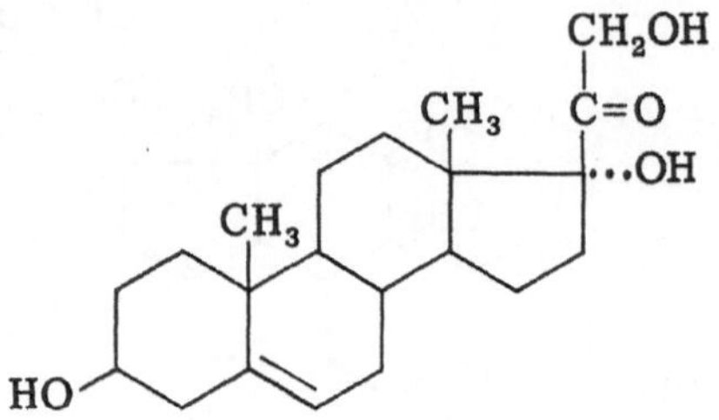

$C_{21}H_{32}O_4$

$3\beta,17\alpha,21$-Trihydroxypregn-5-en-20-one

3-SULFATE

CC, PE ⟨63-11⟩.

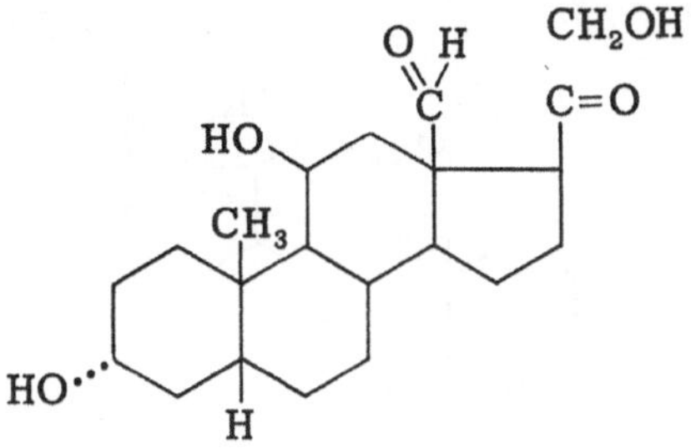

3α,11β,21-Trihydroxy-20-oxo-5β-pregnan-18-al
(Tetrahydroaldosterone)

GLUCURONIDE

CC ⟨64-15⟩ ⟨65-21⟩ • PC ⟨64-15⟩ • PE ⟨64-15⟩.

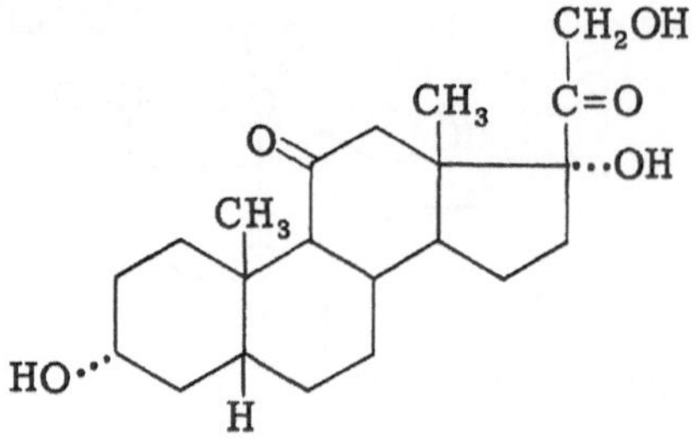

$3\alpha,17\alpha,21$-Trihydroxy-5β-pregnane-11,20-dione
(Tetrahydro-E, Tetrahydrocortisone)

3-β-D-GLUCOPYRANOSIDURONIC ACID

Enzymatic synthesis ⟨56-3⟩.

3-β-D-GLUCOPYRANOSIDURONIC ACID-21-ACETATE

2′,3′,4′-Tri-*O*-acetyl methyl ester — $C_{36}H_{50}O_{15}$ (anal. C, H), mp 209-212°, $[\alpha]_D^{30}$ + 37.8° (chloroform) ⟨55-4⟩ • mp 210-212°, $[\alpha]_D^{20}$ + 32° (chloroform), CC, PC ⟨60-8⟩ ⟨64-6⟩ • mp 210-212°, CC ⟨64-10⟩.

Allied Characterization Data

3-GLUCURONIDE

CC ⟨60-8⟩ ⟨64-6⟩ ⟨64-10⟩ • CCD ⟨55-3⟩ ⟨59-6⟩ • PC ⟨55-2⟩ ⟨57-1⟩ ⟨59-1⟩ ⟨59-9⟩ ⟨63-12⟩ ⟨63-13⟩ ⟨64-6⟩ ⟨64-15⟩ • PE ⟨56-3⟩ ⟨59-1⟩ ⟨64-11⟩.

CH3 / C=O / CH3 / CH3 / HO / H

3 α-Hydroxy-5 α-pregnan-20-one

3-GLUCURONIDE

TLC ⟨64-3⟩.

3-SULFATE

Pyridine salt – $C_{26}H_{39}NO_5S$ (anal. C, H, S), mp 184°, $[\alpha]_D^{24}$ + 70° (chloroform) ⟨60-18⟩.

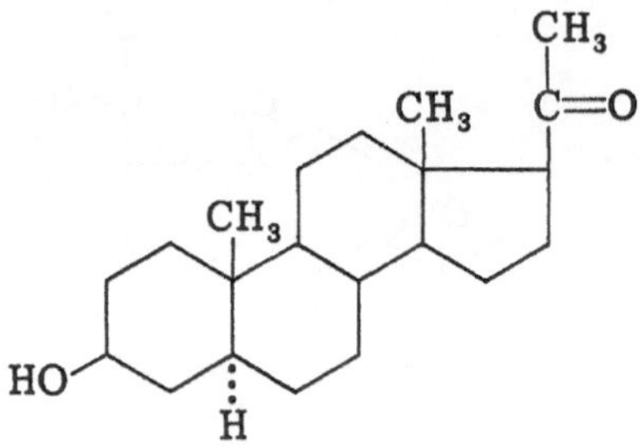

3β-Hydroxy-5α-pregnan-20-one

3-β-D-GLUCOPYRANOSIDURONIC ACID

2′,3′,4′-Tri-*O*-acetyl methyl ester – $C_{34}H_{50}O_{11}$ (anal. C, H), mp 201-202° ⟨60-12⟩.

3-β-D-GLUCOPYRANOSIDURONIC ACID-20-OXIME

2′,3′,4′-Tri-*O*-acetyl methyl ester – $C_{34}H_{51}NO_{11}$ (anal. C, H, N), mp 184-187° ⟨60-12⟩.

3-SULFATE

Methyl ester – $C_{22}H_{36}O_5S$ (anal. C, H, S), mp 136°, $[\alpha]_D^{21}$ + 76° (chloroform) ⟨60-18⟩.

Piperazine salt – mp 240-241° ⟨48-4⟩ • $C_{25}H_{45}N_2O_5S \cdot 2H_2O$ (anal. C, H, N, S), mp 239-241° ⟨48-5⟩.

Potassium salt – mp 207-210° ⟨48-2⟩.

Pyridine salt – $C_{26}H_{39}NO_5S$ (anal. C, H, N), mp 193°, $[\alpha]_D^{24}$ + 67° (chloroform) ⟨60-18⟩ • mp 182-187° ⟨48-5⟩.

Allied Characterization Data

3-SULFATE

CC, PC ⟨66-9⟩.

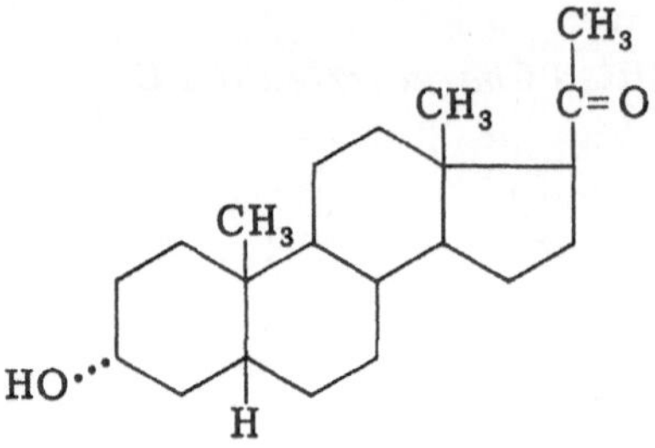

3α-Hydroxy-5β-pregnan-20-one
(Pregnanolone)

3-β-D-GLUCOPYRANOSIDURONIC ACID

Sodium salt – $C_{27}H_{41}NaO_8$ (anal. C, H, Na poor), mp 257-260° ⟨47-2⟩.

Allied Characterization Data

3-GLUCURONIDE

TLC ⟨64-3⟩.

* * * * *

3-SULFATE

Methyl ester – $C_{22}H_{36}O_5S$ (anal. C, H, S), mp 123°, $[\alpha]_D^{24}$ + 108° (chloroform) ⟨60-18⟩.

Potassium salt – mp 232-234° ⟨48-2⟩.

Pyridine salt – $C_{26}H_{39}NO_5S$ (anal. C, H, N), mp 170°, $[\alpha]_D^{24}$ + 103° (chloroform) ⟨60-18⟩.

Triethylammonium salt – $C_{27}H_{49}NO_5S$ (anal. C, H, N, S), mp 224-226°, $[\alpha]_D^{25}$ + 83° (chloroform), NMR ⟨68-2⟩.

Allied Characterization Data

3-SULFATE

CC ⟨65-3⟩⟨66-9⟩ • PC ⟨64-1⟩⟨66-9⟩ • PE ⟨62-9⟩ ⟨64-1⟩.

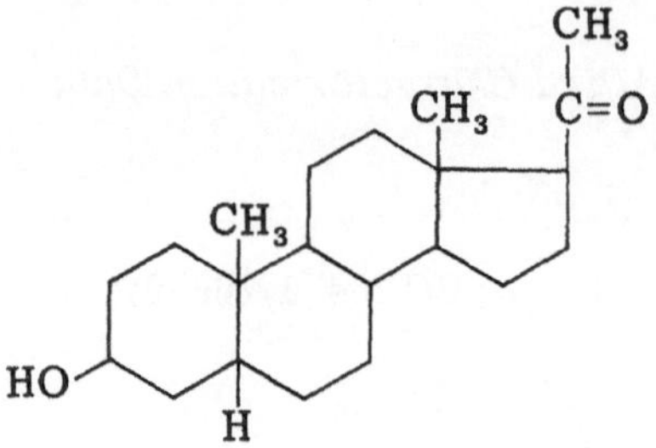

$C_{21}H_{34}O_2$

CH3, CH3, C=O, CH3, HO, H

3β-Hydroxy-5β-pregnan-20-one

3-SULFATE

PC, PE ⟨64-1⟩.

CH3 / CH3 HC–OH / CH3 / HO

Pregn-5-ene-3β,20α-diol

20-2′-AMINO-2′-DEOXY-α-D-GLUCOPYRANOSIDE

2′-Acetyl – $C_{29}H_{47}NO_7 \cdot 2H_2O$ (anal. C, H, N), mp 268-272°, $[\alpha]_D^{25}$ – 17.0° (methanol), CC, IR, ORD ⟨67-2⟩.

2′,3′,4′,6′-Tetraacetyl – $C_{35}H_{53}NO_{10} \cdot 2H_2O$ (anal. C, H, N), mp 252-254°, and acetylated with ^3H-acetic anhydride, CC ⟨67-2⟩.

20-2′-AMINO-2′-DEOXY-α-D-GLUCOPYRANOSIDE-3-SULFATE

2′-Acetyl ammonium salt – $C_{29}H_{51}N_2O_{10}S$, mp 189°, $[\alpha]_D^{25}$ – 19.8° (methanol), IR, NMR ⟨67-2⟩.

2′-Acetyl sodium salt – $C_{29}H_{47}NNaO_{10}S \cdot 3H_2O$ (anal. C, H, N, S), mp 212-214°, CC ⟨67-2⟩.

3-SULFATE

Ammonium salt – mp 196-198°, CC, IR ⟨66-3⟩⟨66-4⟩.

3,20-DISULFATE

Diammonium salt – $C_{21}H_{40}N_2O_8S_2$ (anal. S), mp 198-200°, CC, IR ⟨67-2⟩.

Allied Characterization Data

3-SULFATE

CC ⟨66-5⟩.

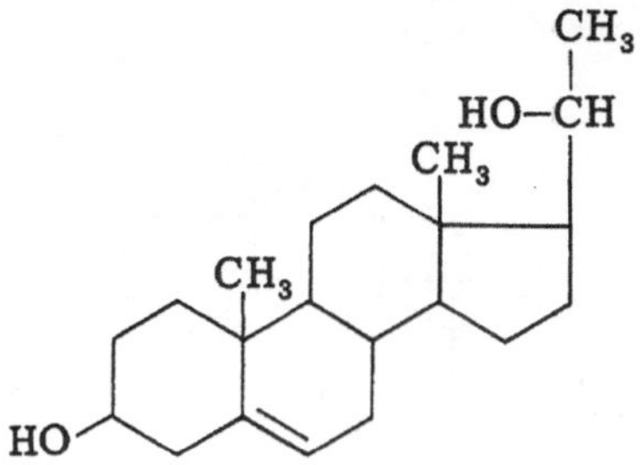

CH_3

HO—CH

CH_3

CH_3

HO

Pregn-5-ene-3β,20β-diol

3-SULFATE

Ammonium salt – mp 196-198°, CC, IR ⟨66-3⟩.

3α,17α-Dihydroxy-5β-pregnan-20-one

3-β-D-GLUCOPYRANOSIDURONIC ACID

Sodium salt – $C_{27}H_{41}NaO_9$ (anal. C, H, Na), mp 266-268° ⟨47-1⟩.

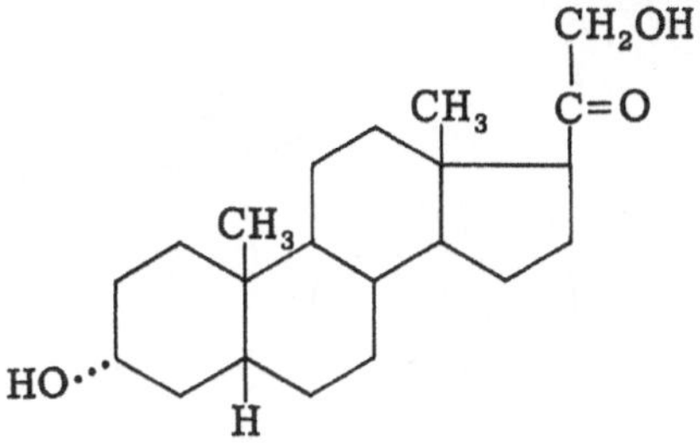

3α,21-Dihydroxy-5β-pregnan-20-one

3-GLUCURONIDE

CC ⟨67-20⟩ • PC ⟨63-13⟩.

21-GLUCURONIDE

CC ⟨67-20⟩.

3,21-DIGLUCURONIDE

CC ⟨67-20⟩.

GLUCURONIDE (Unspecified Location or Number)

CC, PC, PE ⟨61-14⟩.

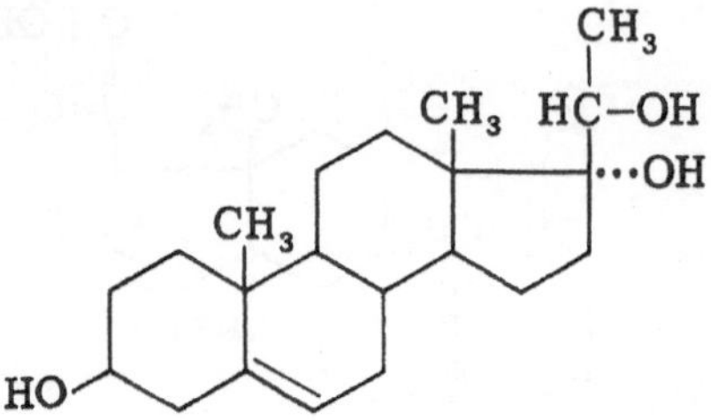

Pregn-5-ene-3β,17α,20α-triol

3-SULFATE

Ammonium salt – mp 179-181°, ^{35}S, 7α-^{3}H, CC, IR ⟨64-21⟩ • mp 187-190°, CC ⟨66-4⟩.

Allied Characterization Data

3-SULFATE

CC ⟨66-4⟩ ⟨66-5⟩ • PC ⟨56-5⟩.

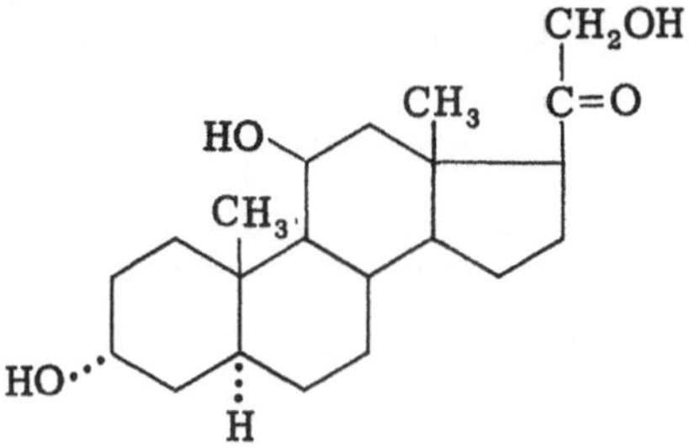

$3\alpha,11\beta,21$-Trihydroxy-5α-pregnan-20-one

3-β-D-GLUCOPYRANOSIDURONIC ACID

Acid — CC ⟨60-8⟩ • CC, PC ⟨64-6⟩.

2′,3′,4′-Tri-*O*-acetyl methyl ester — CC, PC ⟨64-6⟩.

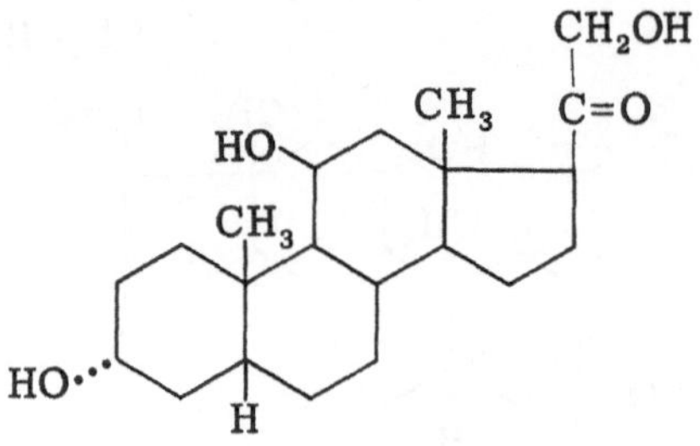

$C_{21}H_{34}O_4$

3α,11β,21-Trihydroxy-5β-pregnan-20-one
(Tetrahydrocorticosterone)

21-SULFATE

^{3}H, ^{35}S, PC ⟨64-18⟩ • CC, PC, PE ⟨65-20⟩.

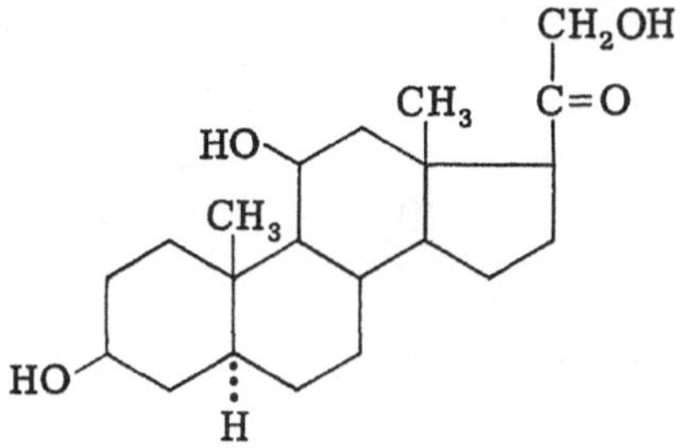

$3\beta,11\beta,21$-Trihydroxy-5α-pregnan-20-one

21-SULFATE

^{3}H, ^{35}S, PC ⟨64-18⟩.

$C_{21}H_{34}O_4$

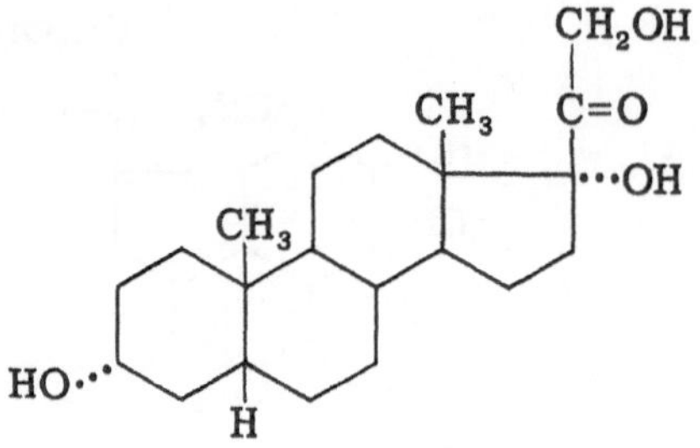

CH₂OH C=O OH CH₃ CH₃ HO H

3α,17α,21-Trihydroxy-5β-pregnan-20-one
(Tetrahydro-S)

3-β-D-GLUCOPYRANOSIDURONIC ACID-21-ACETATE

2′,3′,4′-Tri-*O*-acetyl methyl ester – mp 165-167°, CC, PC ⟨64-6⟩⟨64-10⟩ • CC ⟨60-8⟩.

Allied Characterization Data

3-GLUCURONIDE

PC ⟨63-13⟩⟨64-10⟩⟨64-15⟩.

21-GLUCURONIDE

CC, PC ⟨63-12⟩.

GLUCURONIDE (Unspecified Location)

CC ⟨60-8⟩.

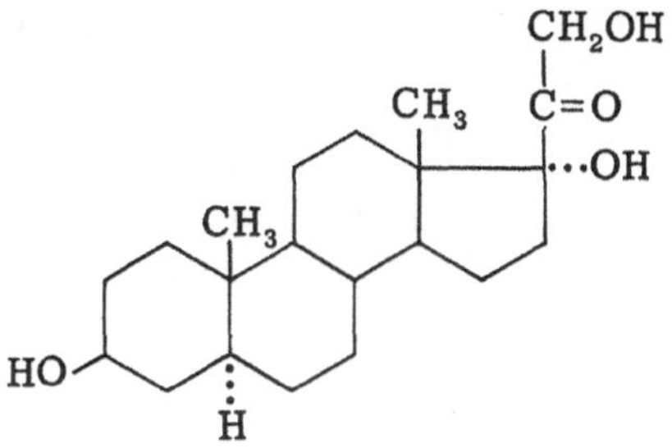

3β,17α,21-Trihydroxy-5α-pregnan-20-one

3,21-DI-β-D-GLUCOPYRANOSIDURONIC ACID

Di-2′,3′,4′-Tri-*O*-acetyl methyl ester — $C_{47}H_{66}O_{22}$ (anal. C, H), mp 176-177°, $[\alpha]_D^{28}$ − 15.3° (chloroform) ⟨57-11⟩⟨59-7⟩ • IR ⟨59-8⟩.

HO, CH_3, CH_3, CH_2OH, C=O, OH, HO, H

3α,11β,17α,21-Tetrahydroxy-5β-pregnan-20-one
(Tetrahydro-F, Tetrahydrohydrocortisone, Tetrahydrocortisol)

3-β-D-GLUCOPYRANOSIDURONIC ACID

Enzymatic synthesis ⟨56-3⟩.

3-β-D-GLUCOPYRANOSIDURONIC ACID-21-ACETATE

2′,3′,4′-Tri-*O*-acetyl methyl ester — CC, PC ⟨64-6⟩ • CC ⟨60-8⟩.

Allied Characterization Data

3-GLUCURONIDE

CC ⟨60-8⟩ ⟨64-6⟩ • PC ⟨57-1⟩ ⟨59-1⟩ ⟨59-9⟩ ⟨63-12⟩ ⟨63-13⟩ ⟨64-6⟩ ⟨64-15⟩ • PE ⟨59-1⟩.

* * * * *

3-SULFATE

Sodium salt — $C_{21}H_{33}NaO_8S$ (anal. C, H, S), mp 142-145°, CC, PE ⟨64-12⟩.

3,21-DISULFATE

Disodium salt – $C_{21}H_{32}Na_2O_{11}S_2$ (anal. C, H, S), mp 135-136°, CC, PE ⟨64-12⟩.

3,17,21-TRISULFATE

Trisodium salt – $C_{21}H_{31}Na_3O_{14}S_3$ (anal. C, H, S), CC, PE ⟨64-11⟩⟨64-12⟩.

Allied Characterization Data

21-SULFATE

PC ⟨64-18⟩ • PE ⟨64-11⟩.

3,21-DISULFATE

PE ⟨64-11⟩.

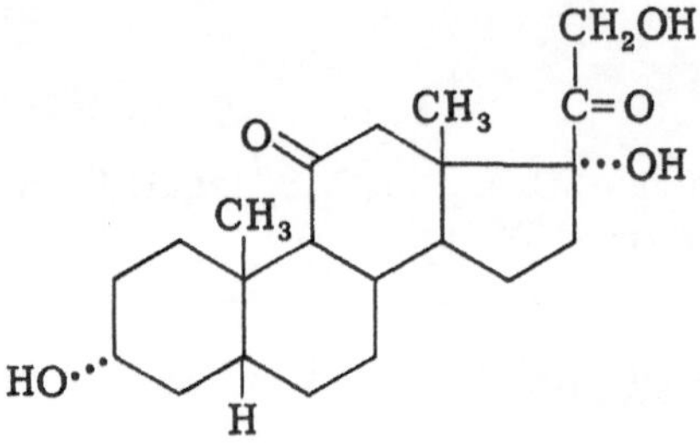

$3\alpha,17\alpha,20\alpha,21$-Tetrahydroxy-$5\beta$-pregnan-11-one
(Cortolone)

GLUCURONIDE (Unspecified Location)

CC ⟨55-1⟩.

GLUCURONIDE TRI-*O*-ACETYL METHYL ESTER

CC ⟨60-8⟩.

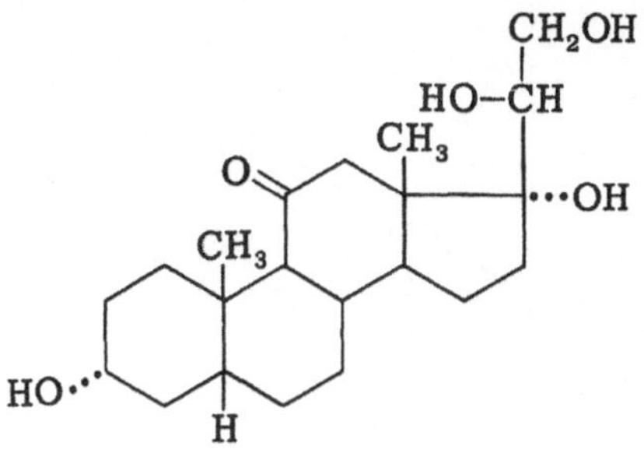

3α,17α,20β,21-Tetrahydroxy-5β-pregnan-11-one
(β-Cortolone)

GLUCURONIDE (Unspecified Location)

CC ⟨55-1⟩ • CCD ⟨59-6⟩ • PC ⟨55-2⟩.

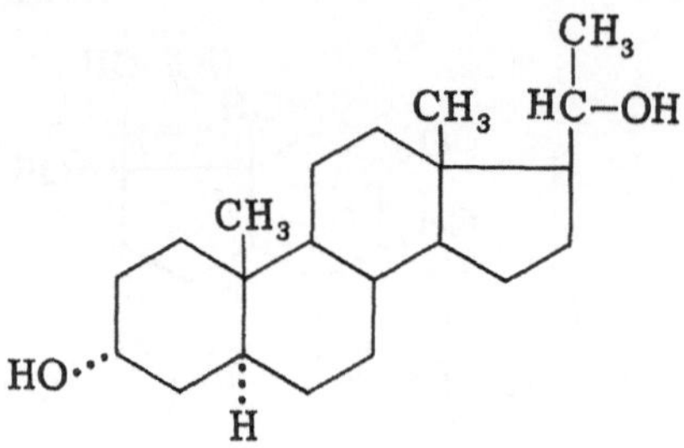

5α-Pregnane-3α,20α-diol

3-GLUCURONIDE

TLC ⟨64-3⟩.

SULFATE (Unspecified Location)

TLC ⟨67-12⟩.

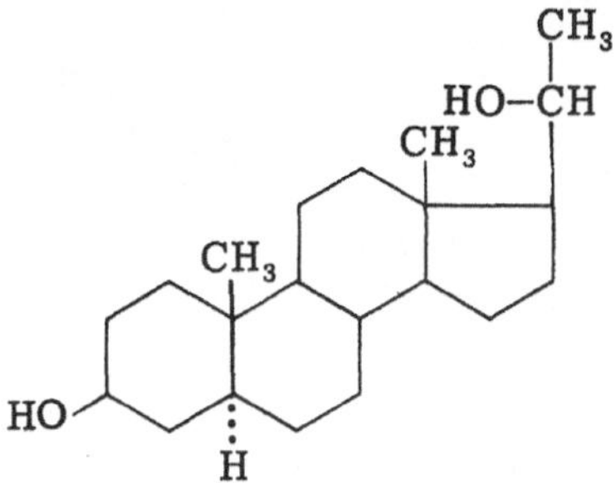

5α-Pregnane-3β,20β-diol

20-SULFATE

Potassium salt – $C_{21}H_{35}KO_5S \cdot H_2O$ (anal. C, H, K), mp 229-231°, $[\alpha]_D^{22}$ + 22° (water), IR ⟨63-9⟩.

Toluidine salt – $C_{28}H_{45}NO_5S$ (anal. C, H), mp 178-183°, $[\alpha]_D^{23}$ + 18° (ethanol) ⟨63-9⟩.

20-SULFATE-3-ACETATE

Potassium salt - mp 246-250°, IR ⟨63-9⟩.

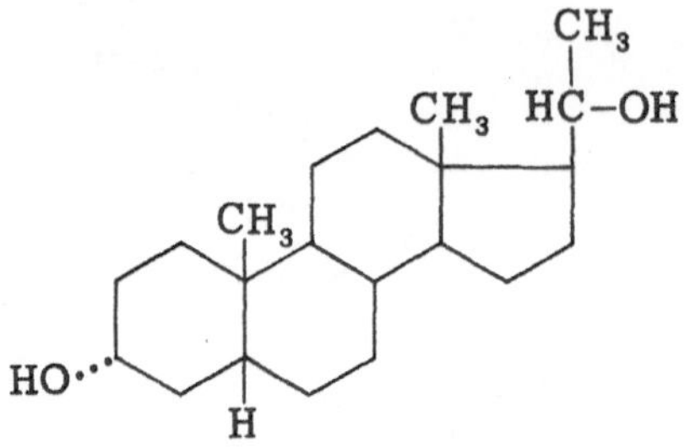

5β-Pregnane-3α,20α-diol
(Pregnanediol)

3-β-D-GLUCOPYRANOSIDURONIC ACID

Acid – $C_{27}H_{44}O_8$ (anal. C, H), mp 178-180°, $[\alpha]_D^{20.5}$ – 5.0° (ethanol) ⟨47-2⟩ • $C_{27}H_{44}O_8 \cdot H_2O$ (anal. C, H), mp 179-180° ⟨36-3⟩ • mp 178-180° ⟨44-2⟩ • mp 171-172° ⟨44-3⟩.

Barium salt – $C_{27}H_{43}Ba_{1/2}O_8$ (anal. C, H, Ba), mp 272° ⟨44-3⟩.

Calcium salt – $C_{27}H_{43}Ca_{1/2}O_8$, mp 273-274° ⟨44-3⟩.

Sodium salt – $C_{27}H_{43}NaO_8 \cdot 1\frac{1}{2}H_2O$ (anal. C, H, Na), mp 270-272°, $[\alpha]_D^{19}$ – 11° (aqueous ethanol) ⟨42-4⟩ • $C_{27}H_{43}NaO_8 \cdot 2H_2O$ (anal. C, H, Na), mp 271-273°, $[\alpha]_D^{20}$ – 9.1° (aqueous ethanol) ⟨42-5⟩ • $C_{27}H_{43}NaO_8 \cdot 3H_2O$ (anal. C, H, Na), mp 283.5-284.5° ⟨47-2⟩ • mp 274-275° ⟨44-2⟩ • isolated ⟨36-3⟩ • mp 264-270° ⟨44-3⟩.

Strontium salt — mp 272° ⟨44-3⟩.

3-β-D-GLUCOPYRANOSIDURONIC ACID-20-ACETATE

γ-Lactone-2′,4′-diacetyl – $C_{33}H_{48}O_{10}$ (anal. C, H, acetyl), mp 123-125° ⟨44-1⟩.

2′,3′,4′-Tri-*O*-acetyl methyl ester – $C_{36}H_{54}O_{12}$ (anal. C, H, OCH_3), mp 191-192°, $[\alpha]_D^{30}$ + 7.5° (benzene) ⟨44-2⟩.

Allied Characterization Data

3-GLUCURONIDE

CC ⟨62-8⟩⟨65-3⟩ • PC ⟨55-2⟩⟨62-8⟩ • PE ⟨64-11⟩ • TLC ⟨64-3⟩.

MONOGLUCURONIDE (Unspecified Location)

PE ⟨57-2⟩ • CC ⟨66-23⟩ • PC ⟨57-2⟩⟨66-23⟩ • TLC ⟨66-37⟩.

MONOGLUCURONIDE TRI-*O*-METHYLSILYL METHYLESTER (Unspecified Location)

GLC ⟨67-16⟩.

* * * * *

3-SULFATE

CC ⟨65-3⟩.

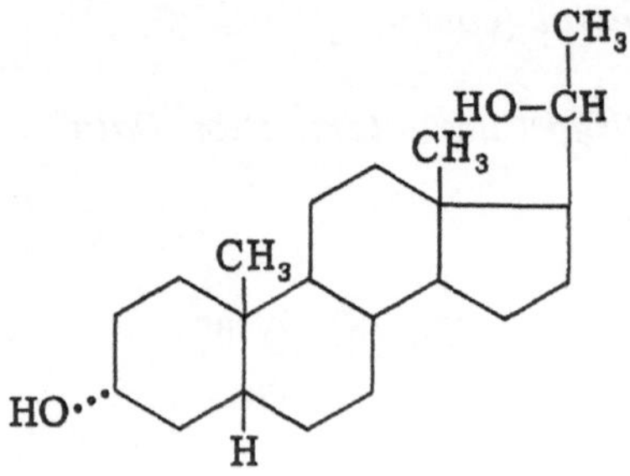

5β-Pregnane-3α,20β-diol

3,20-DISULFATE

Methyl ester – $C_{23}H_{40}O_8S_2$ (anal. C, H, S), mp 106°, $[\alpha]_D^{21}$ + 21° (chloroform) ⟨60-18⟩.

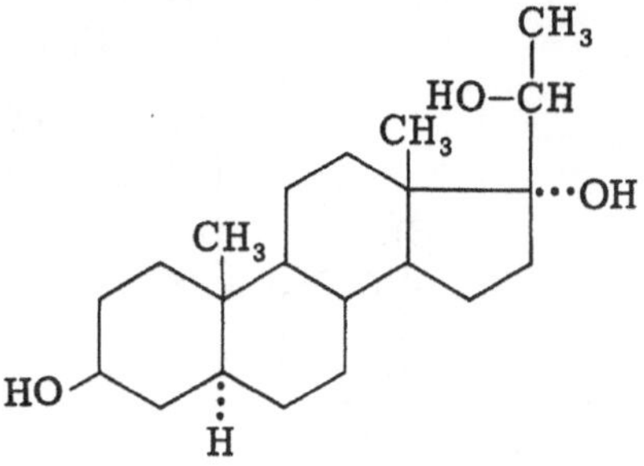

5α-Pregnane-3β,17α,20β-triol

20-SULFATE

Sodium salt – $C_{21}H_{35}NaO_6S$ (anal. C, H, S), mp 133-141° ⟨65-27⟩.

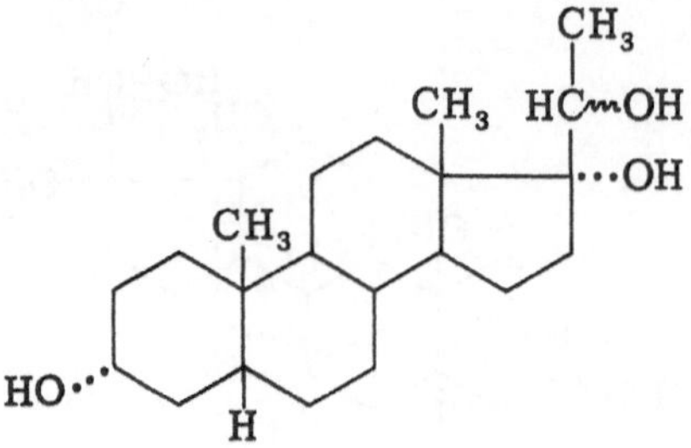

5β-Pregnane-3α,17α,20ξ-triol

β-D-GLUCOPYRANOSIDURONIC ACID (Unspecified Location)

Sodium salt – ⟨45-1⟩.

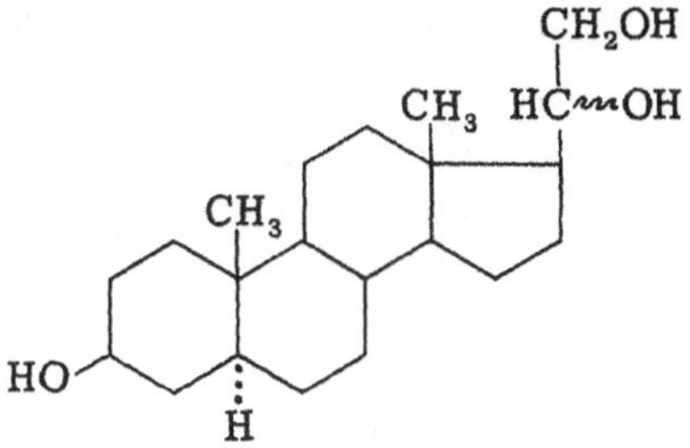

5α-Pregnane-3β,20ξ,21-triol

SULFATE (Unspecified Location)

Enzymatic synthesis — PC ⟨56-5⟩.

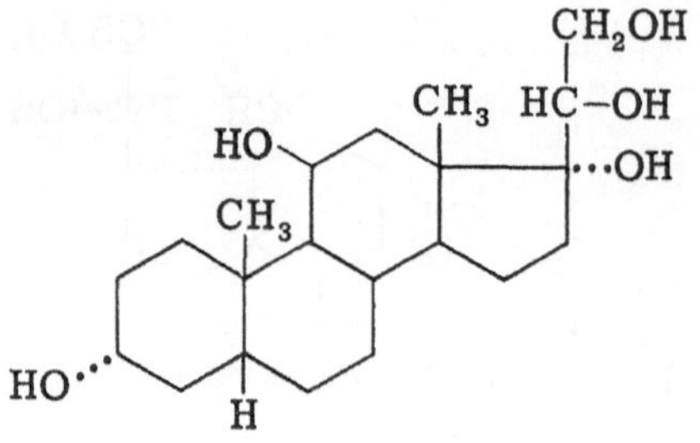

5β-Pregnane-3α,11β,17α,20α,21-pentol
(Cortol)

GLUCURONIDE (Unspecified Location)

⟨55-1⟩.

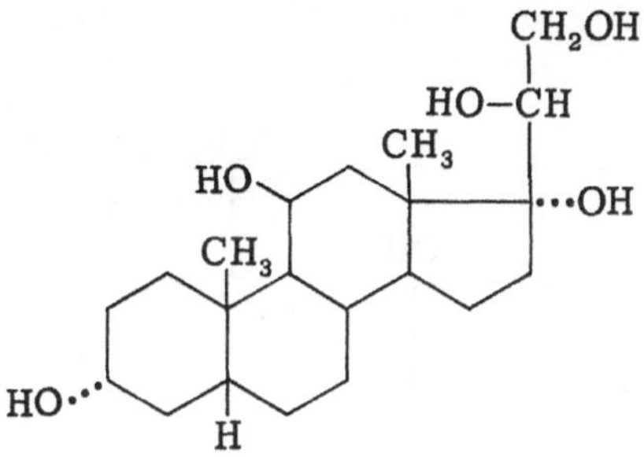

5β-Pregnane-3α,11β,17α,20β,21-pentol
(β-Cortol)

GLUCURONIDE (Unspecified Location)

⟨55-1⟩.

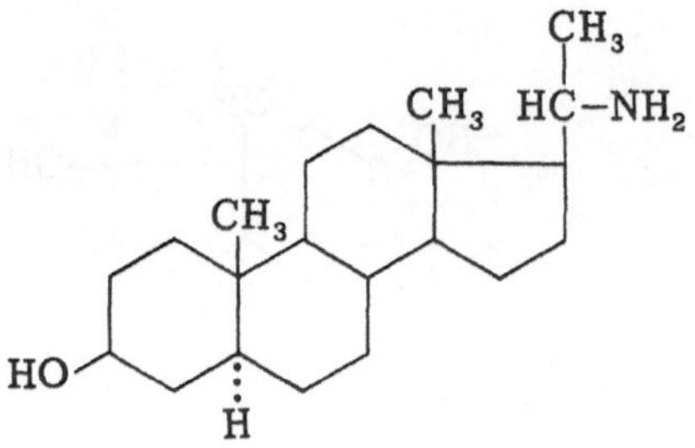

20α-Amino-5α-pregnan-3β-ol

3-β-D-GLUCOPYRANOSIDURONIC ACID

Sodium salt – $C_{21}H_{44}NNaO_7$ (anal. C, H, N), mp > 300° ⟨60-12⟩.

2′,3′,4′-Tri-*O*-acetyl methyl ester – $C_{34}H_{53}NO_{10}$ (anal. C, H), mp 178-179° ⟨60-12⟩.

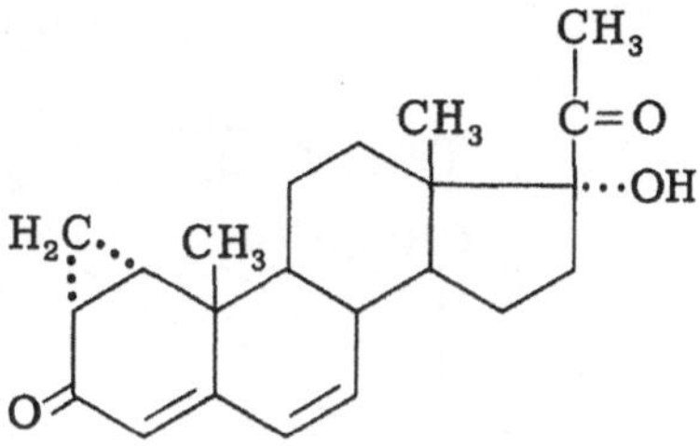

17α-Hydroxy-1α,2α-methanopregna-4,6-diene-3,20-dione

17-SULFATE

Sodium salt – mp 152° ⟨63-6⟩ ⟨64-23⟩.

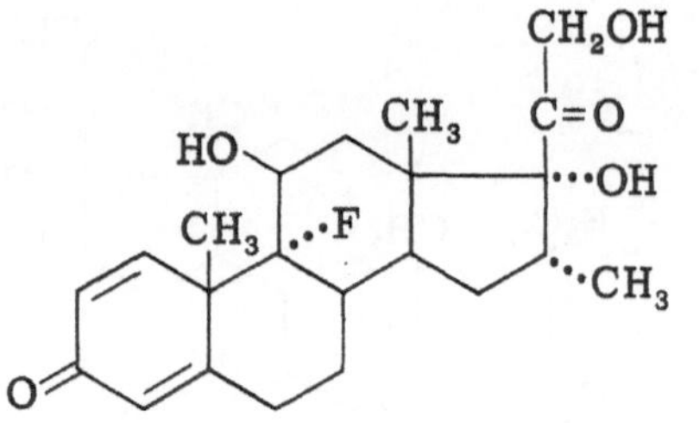

9α-Fluoro-11β,17α,21-trihydroxy-16α-methylpregna-1,4-diene-3,20-dione
(Dexamethasone)

21-SULFATE

Sodium salt – $C_{22}H_{28}FNaO_8S$, mp 207-209°, $[\alpha]_D^{23}$ + 84° (water), $\lambda_{max}^{methanol}$ 239 mμ (ε 15,100), IR ⟨65-19⟩ • mp 182-182.5°, $[\alpha]_D$ + 64° (water), pH 7.0 (aqueous solution) ⟨60-9⟩⟨64-7⟩ • $[\alpha]_D$ 80-85° (water), λ_{max} 238 mμ ($E_{1cm}^{1\%}$ 292) ⟨65-26⟩.

Triethylammonium salt – $C_{28}H_{44}FNO_8S$, mp 199-201°, λ_{max} 238 mμ (ε 14,920) ⟨60-9⟩⟨64-7⟩.

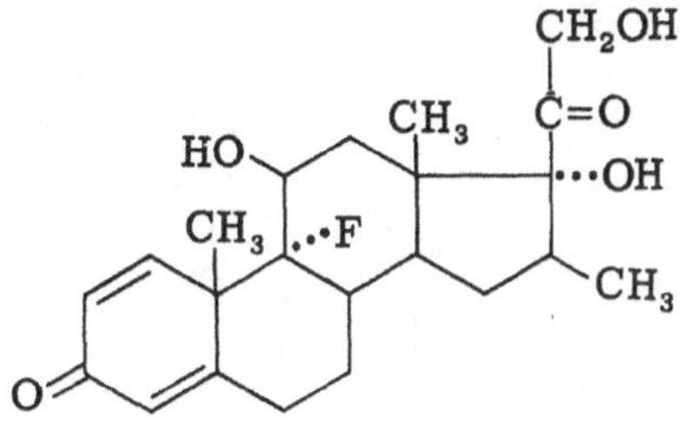

9α-Fluoro-11β,17α,21-trihydroxy-16β-methylpregna-1,4-diene-3,20-dione
(Betamethasone)

21-SULFATE

Sodium salt – $[\alpha]_D$ 90-95° (water), λ_{max} 238 mμ ($E^{1\%}_{1cm}$ 291) ⟨65-26⟩.

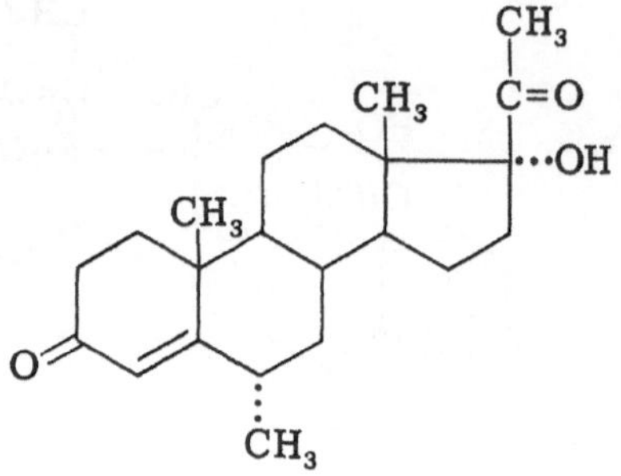

17α-Hydroxy-6α-methylpregn-4-ene-3,20-dione

17-SULFATE

Sodium salt – $C_{22}H_{31}NaO_6S$ (anal. Na, S), IR ⟨63-6⟩ ⟨64-23⟩.

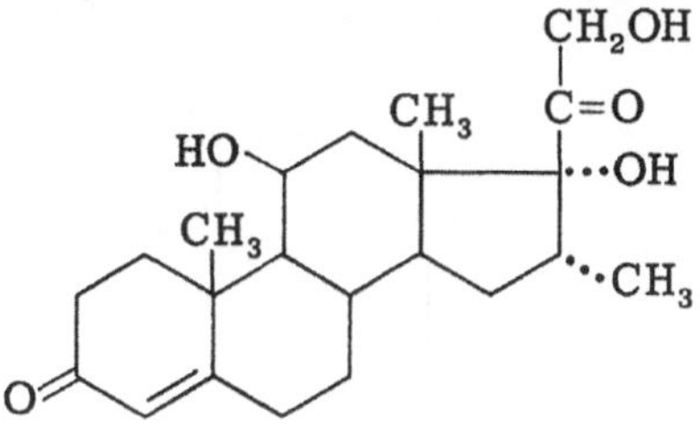

11β,17α,21-Trihydroxy-16α-methylpregn-4-ene-3,20-dione

21-β-D-GLUCOPYRANOSIDURONIC ACID

2′,3′,4′-Tri-*O*-acetyl methyl ester – λ_{max} 240 mμ ⟨63-15⟩.

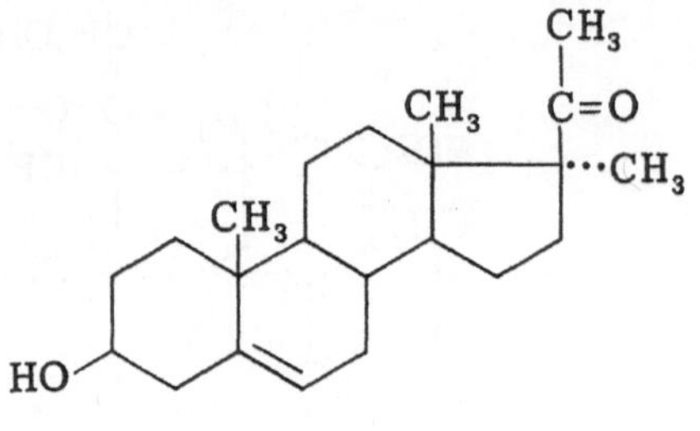

3β-Hydroxy-17α-methylpregn-5-en-20-one

3-SULFATE

Ammonium salt – mp 225-227° ⟨66-1⟩ • no physical constants ⟨65-6⟩.

Pyridine salt – no physical constants ⟨66-1⟩.

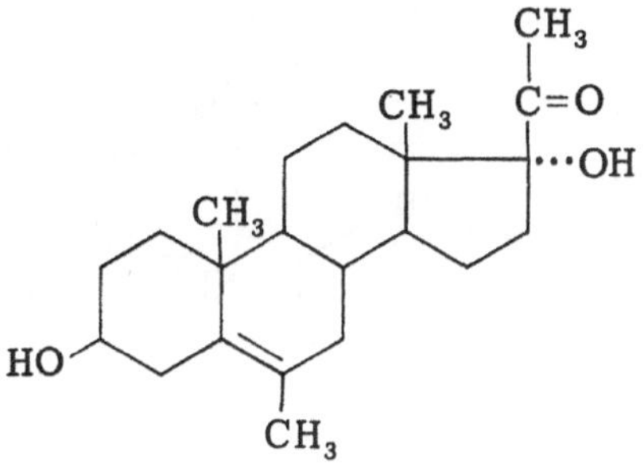

3β,17α-Dihydroxy-6-methylpregn-5-en-20-one

3-SULFATE-17-ACETATE

Ammonium salt – no physical constants ⟨65-6⟩.

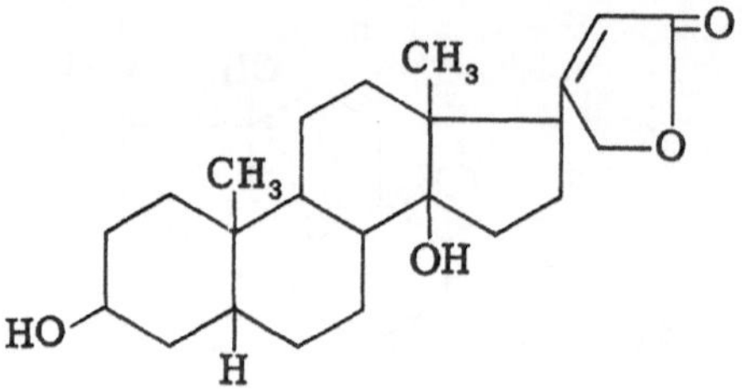

3β,14β-Dihydroxy-5β-card-20(22)-enolide
(Digitoxigenin)

3-β-D-GLUCOPYRANOSIDURONIC ACID

2′,3′,4′-Tri-*O*-acetyl methyl ester — $C_{36}H_{50}O_{13}$ (anal. C, H), mp 225.5-228°, $[\alpha]_D^{20}$ + 2.1° (methanol) ⟨62-16⟩.

H_3C CH_3 CH_3 CH_3 CH_3 O

Cholest-4-en-3-one
(Cholestenone)

3,5-DIEN-3-OL-β-D-GLUCOPYRANOSIDURONIC ACID

2′,3′,4′-Tri-*O*-acetyl methyl ester – $C_{40}H_{60}O_{10}$ (anal. C, H), mp 156-158°, $[\alpha]_D^{23}$ – 55.3° (chloroform) ⟨59-7⟩ • IR ⟨59-8⟩.

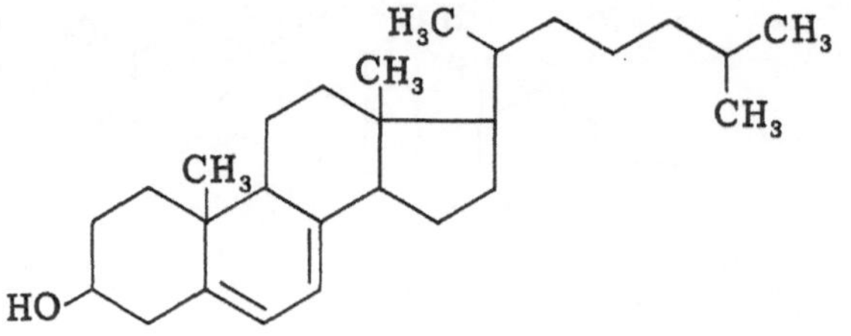

Cholesta-5,7-dien-3β-ol
(7-Dehydrocholesterol)

3-SULFATE

Ammonium salt – $C_{27}H_{47}NO_4S \cdot \frac{1}{2}H_2O$ (anal. C, H, N, S, H_2O), mp 176.5-177°, $[\alpha]_D^{25}$ – 49° (dimethylformamide), $\lambda_{max}^{methanol}$ 261 (shoulder), 271, 282 and 294 mμ (ϵ 7000, 10,250, 10,830 and 6140) ⟨68-6⟩.

Sodium salt – $C_{27}H_{43}NaO_4S \cdot H_2O$ (anal. C, H, Na, S, H_2O), mp 137.5-138.5°, $[\alpha]_D^{25}$ – 48° (dimethylformamide), $\lambda_{max}^{methanol}$ 262 (shoulder), 271, 282, 294 mμ (ϵ 7950, 11,350, 11,850 and 6940) ⟨68-6⟩.

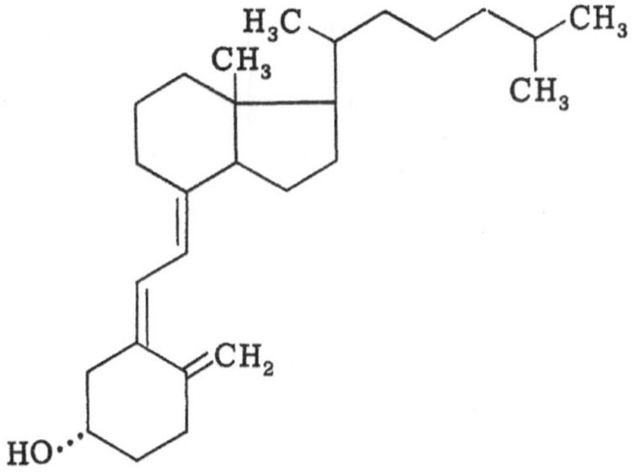

Vitamin D_3

3-SULFATE

Ammonium salt – $C_{27}H_{47}NO_4S \cdot \frac{1}{4}H_2O$ (anal. C, H, N, S, H_2O), amorphous solid, $[\alpha]^{25}_{5461}$ + 51° (chloroform), $\lambda^{methanol}_{max}$ 212 and 266 mμ (ε 15,600, 17,650) ⟨68-6⟩.

Barium salt – ⟨67-27⟩.

Potassium salt – amorphous solid, mp 124-134°, $\lambda^{methanol}_{max}$ 212 and 266 mμ (ε 13,000, and 14,200) ⟨68-6⟩.

Pyridine salt – $C_{32}H_{49}NO_4S$ (anal. C, H, N, S), mp 127-131°, $[\alpha]^{25}_D$ + 51° (chloroform), $\lambda^{methanol}_{max}$ 212, 251 (shoulder), 257 and 263 mμ (ε 16,650, 17,500, 19,170, 19,500) ⟨68-6⟩.

Sodium salt – amorphous solid, mp 113-120°, $\lambda^{methanol}_{max}$ 212 and 266 mμ (ε 14,300, 16,000) ⟨68-6⟩.

Allied Characterization Data

3-SULFATE

TLC ⟨67-27⟩.

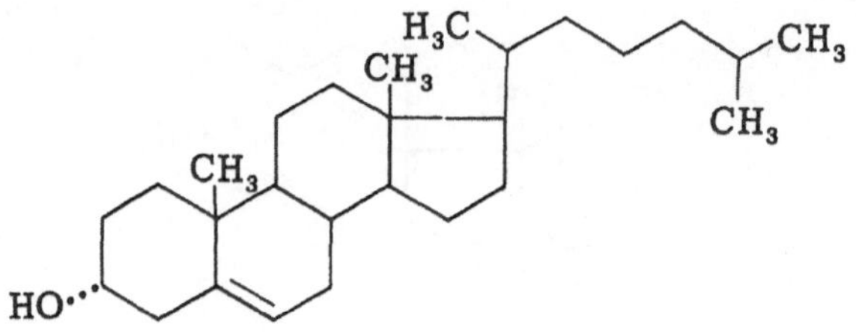

Cholest-5-en-3α-ol
(Epicholesterol)

3-SULFATE

Pyridine salt – $C_{32}H_{51}NO_4S$ (anal. S), mp 163-164° ⟨52-1⟩.

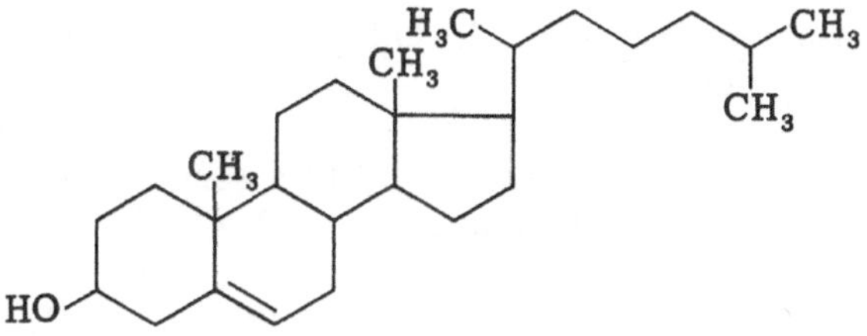

Cholest-5-en-3β-ol
(Cholesterol)

3-β-D-GALACTOPYRANOSIDURONIC ACID

2′,3′,4′-Tri-*O*-acetyl methyl ester – $C_{40}H_{62}O_{10}$ (anal. C, H, OCH_3), mp 219-220°, $[\alpha]_D^{27.5}$ – 6.36° (chloroform) ⟨38-2⟩.

3-β-D-GLUCOPYRANOSIDURONIC ACID

Acid – $C_{33}H_{54}O_7$ (anal. C, H, O), mp 242-245°, IR, TLC ⟨66-31⟩ • mp 272-274°, IR, PC, TLC ⟨61-17⟩.

Barium salt – No details given ⟨59-3⟩.

Cinchonine salt – $C_{19}H_{22}N_2O \cdot C_{33}H_{54}O_7$, mp 237-242°, PC, TLC ⟨61-17⟩.

Sodium salt – IR ⟨66-31⟩.

2′,3′,4′-Tri-*O*-acetyl methyl ester – $C_{40}H_{62}O_{10}$ (anal. C, H, O), mp 188-190° ⟨59-3⟩ • (anal. C, H, O), mp 176-178°, IR, TLC ⟨66-31⟩ • (anal. C, H), mp 162-164.5° ⟨39-3⟩ • CC ⟨67-6⟩.

3-SULFATE

Ammonium salt – $C_{27}H_{49}NO_4S \cdot \frac{1}{2}H_2O$ (anal. C, H, N, S, H_2O), mp 214-215°, $[\alpha]_D^{25}$ – 24° (dimethylformamide – methanol) ⟨66-25⟩ • mp 185° ⟨15-1⟩ • mp 197-201°, CC ⟨67-9⟩.

Barium salt – $(C_{27}H_{45}O_4S)_2Ba \cdot 3H_2O$ (anal. Ba), mp 124° ⟨41-1⟩.

Calcium salt – $(C_{27}H_{45}O_4S)_2Ca$ (anal. Ca), mp 136° ⟨41-1⟩ • 5,6-Dibromo (anal. Ca) ⟨41-1⟩.

Cupric salt – mp 150° decomposes ⟨41-1⟩.

Lead salt – $(C_{27}H_{45}O_4S)_2Pb$ (anal. Pb), mp 132-134° ⟨41-1⟩.

Lithium salt – $C_{27}H_{45}LiO_4S$, (anal. Li, S), mp 150° ⟨34-1⟩.

Magnesium salt – $(C_{27}H_{45}O_4S)_2Mg \cdot 6H_2O$ (anal. Mg, S), mp 152-154° ⟨41-1⟩.

Mercuric acetate salt – $C_{29}H_{48}HgO_6S$ (anal. Hg), mp 152-171° ⟨41-1⟩ • 5,6-dibromo derivative (anal. Hg), mp 123° ⟨41-1⟩.

Methyl ester – $C_{28}H_{48}O_4S$ (anal. C, H, S), mp 103-104°, $[\alpha]_D$ – 32° ⟨57-8⟩.

N-Methylpyridine salt – $C_{33}H_{53}NO_4S$ (anal. N, S), mp 165-167.5°, $[\alpha]_D$ – 28° ⟨57-8⟩.

Potassium salt – $C_{27}H_{45}KO_4S$ (anal. K, S), mp 212° ⟨34-1⟩ ⟨39-2⟩ • (anal. K, S), mp 226-227° ⟨57-7⟩ • (anal. K, S), mp 225-226° ⟨57-8⟩ • $C_{27}H_{45}KO_4S \cdot H_2O$ (anal. C, H, S), mp 247-248° ⟨66-30⟩ • (anal. K), mp 210° ⟨41-1⟩ • $C_{27}H_{45}KO_4S \cdot 1\frac{1}{4}H_2O$ (anal. C, H, K, S, H_2O), mp 218-220°, $[\alpha]_D^{25}$ – 21° (dimethylformamide – methanol) ⟨66-25⟩ • mp 235° ⟨15-1⟩ • 5,6-Dibromo (anal. K) ⟨41-1⟩.

Pyridine salt – $C_{32}H_{51}NO_4S$ (anal. N, S), mp 179°, $[\alpha]_D^{32}$ – 24° (chloroform) ⟨41-1⟩ • $C_{32}H_{51}NO_4S \cdot CHCl_3$ (anal. S), mp 178° ⟨61-9⟩ • $C_{32}H_{51}NO_4S \cdot CHCl_3$, mp 158-160°, $[\alpha]_D$ – 27° ⟨57-7⟩ • CC ⟨66-41⟩ • 5,6-Dibromo (anal. C_5H_5NH, Br, S), mp 135° ⟨41-1⟩.

Silver salt – $C_{27}H_{45}AgO_4S$ (anal. Ag), mp 124° ⟨41-1⟩.

Sodium salt – $C_{27}H_{45}NaO_4S$ (anal. Na, S), mp 174° ⟨34-1⟩ •

$C_{27}H_{45}NaO_4S \cdot H_2O$ (anal. C, H, Na, S, H_2O), mp 194-195°, $[\alpha]_D^{25}$ − 19° (dimethylformamide − methanol) ⟨66-25⟩ • $C_{27}H_{45}NaO_4S \cdot 3H_2O$ (anal. Na), mp 182-183° ⟨48-3⟩ • $C_{27}H_{45}NaO_4S \cdot 6H_2O$ (anal. Na), mp 177-178.5° ⟨41-1⟩ • mp 184-185° ⟨57-8⟩ • mp 163-164°, $[\alpha]_D^{25}$ − 3° (80% alcohol) ⟨15-1⟩ • 26-^{14}C ⟨66-14⟩ • CC ⟨66-41⟩ • 5,6-Dibromo (anal. Na) ⟨41-1⟩.

Triethylammonium salt − $C_{33}H_{61}NO_4S$ (anal. C, H, N, S), mp 172-175°, $[\alpha]_D^{25}$ − 26.4° (chloroform) ⟨68-3⟩.

Allied Characterization Data

3-SULFATE

CC ⟨64-21⟩⟨66-5⟩ • TLC ⟨66-31⟩.

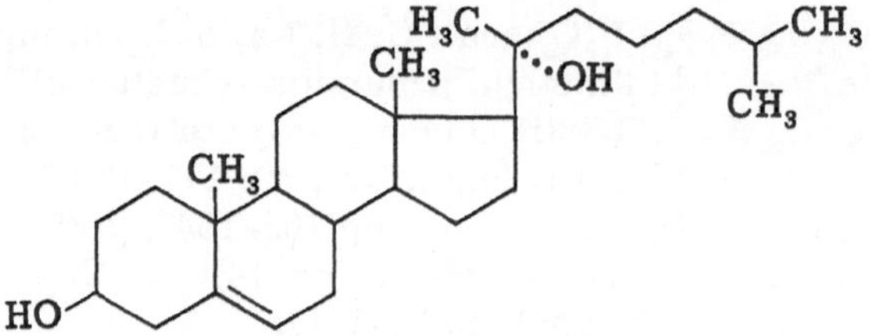

Cholest-5-ene-3β,20α-diol
(20α-Hydroxycholesterol)

3-SULFATE

CC ⟨66-5⟩.

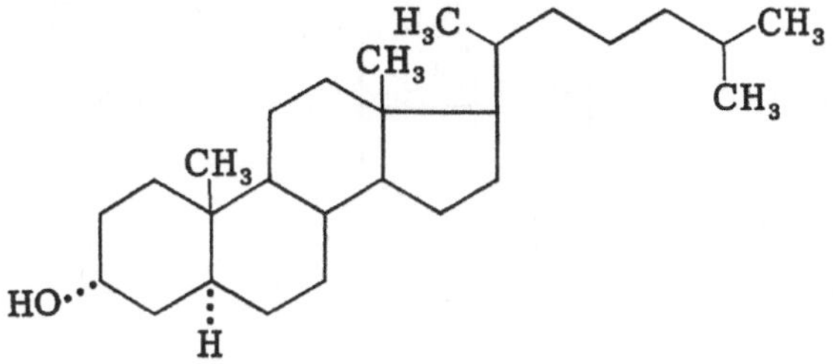

5 α-Cholestan-3 α-ol

3-SULFATE

Sodium salt – $C_{27}H_{47}NaO_4S$ (anal. C, H, Na, S), mp 136-137°, $[\alpha]_D^{29}$ + 15.0° (ethanol) ⟨48-3⟩.

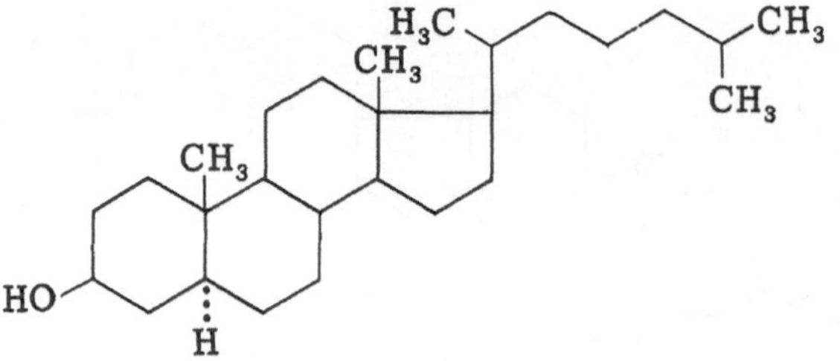

5α-Cholestan-3β-ol
(Cholestanol)

3-SULFATE

Methyl ester – $C_{28}H_{50}O_4S$ (anal. C, H, S), mp 115.5-116.5°, $[\alpha]_D$ + 21° (chloroform) ⟨57-8⟩.

N-Methylpyridine salt – $C_{33}H_{55}NSO_4$ (anal. N, S), mp 175-177°, $[\alpha]_D$ + 19° (chloroform) ⟨57-8⟩.

Potassium salt – $C_{27}H_{47}KO_4S$ (anal. K), mp 234-235° ⟨57-7⟩ • mp 234° ⟨57-8⟩.

Pyridine salt – $C_{32}H_{53}NO_4S \cdot CHCl_3$ (anal. C, H, N, S), mp 165-169° (178-180° in capillary), $[\alpha]_D$ + 17° (chloroform) ⟨57-7⟩.

Sodium salt – $C_{27}H_{47}NaO_4S$ (anal. C, H, Na, S), mp 174-175.5°, $[\alpha]_D^{29}$ + 16.6° (ethanol) ⟨48-3⟩ • mp 173-175° ⟨57-8⟩.

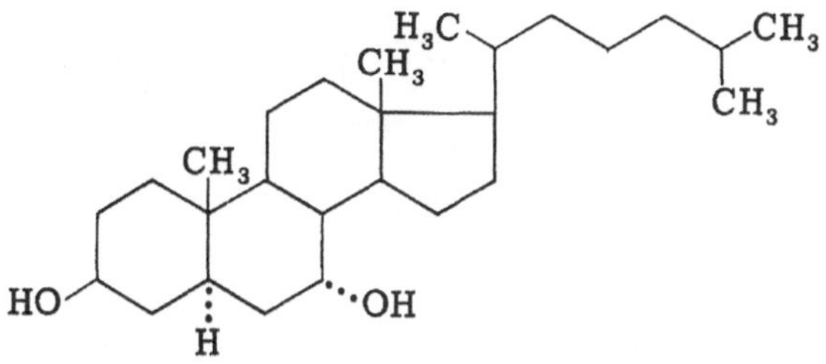

5α-Cholestane-3β,7α-diol

3,7-DISULFATE

Dipotassium salt – $C_{27}H_{46}K_2O_8S_2$ (anal. K), mp 132-135° ⟨54-1⟩.

Dipyridine salt – $C_{37}H_{58}N_2O_8S_2$ (anal. C, H, N, S), mp 132-135°, $\lambda_{max}^{alcohol}$ 256 mμ (ϵ 8050) ⟨54-1⟩.

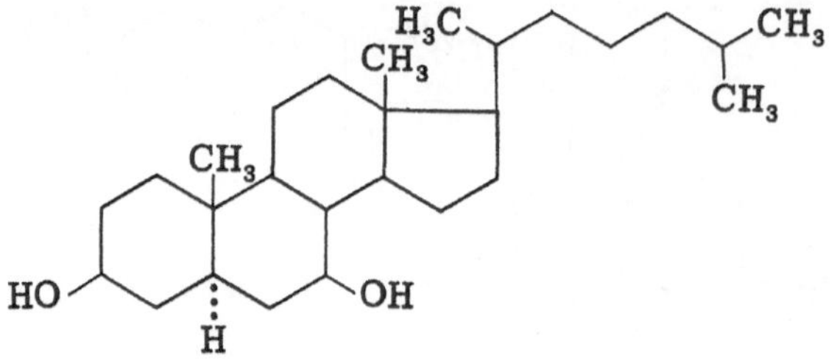

5α-Cholestane-3β,7β-diol

3,7-DISULFATE

Dipotassium salt – $C_{27}H_{46}K_2O_8S_2$ (anal. K), mp 178-181° ⟨54-1⟩.

Dipyridine salt – $C_{37}H_{58}N_2O_8S_2$ (anal. C, H, N, S), mp 157-162°, $[\alpha]_D^{25}$ + 48° (chloroform), $\lambda_{max}^{alcohol}$ 256 mμ (ε 7000) ⟨54-1⟩.

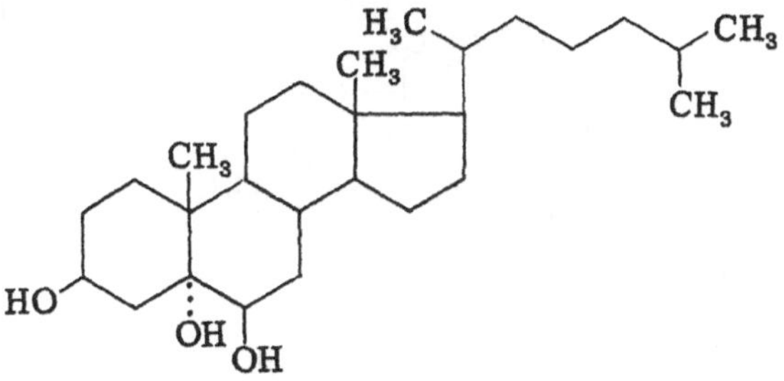

Cholestane-3β,5α,6β-triol

3-SULFATE

Ammonium salt ⟨66-29⟩.

Sodium salt ⟨66-29⟩.

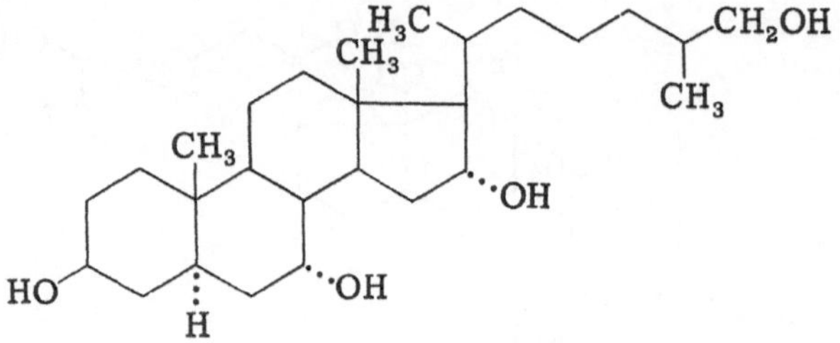

5α-Cholestane-3β,7α,16α,26-tetrol
(Myxinol)

3,26-DISULFATE

Dibarium salt – $C_{27}H_{46}BaO_{10}S_2 \cdot 2H_2O$, mp 147-148° ⟨66-24⟩ ⟨67-1⟩.

Disodium salt – $C_{27}H_{46}Na_2O_{10}S_2$ (anal. Na, S), mp 195° ⟨66-24⟩ ⟨67-1⟩.

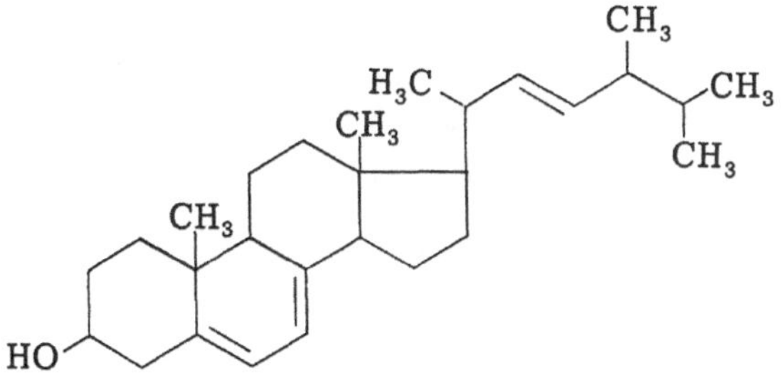

Ergosta-5,7,22-trien-3β-ol
(Ergosterol)

3-β-D-GALACTOPYRANOSIDURONIC ACID

2′,3′,4′-Tri-*O*-acetyl methyl ester – $C_{41}H_{60}O_{10}$ (anal. C, H, OCH_3), mp 204-205°, $[\alpha]_D^{25}$ – 27.9° (chloroform) ⟨38-2⟩.

3-SULFATE

Barium salt – mp 145° ⟨41-1⟩.

Calcium salt – $(C_{28}H_{43}O_4S)_2Ca \cdot 5H_2O$ (anal. Ca, S), mp 135° ⟨41-1⟩.

Lithium salt – $C_{28}H_{43}LiO_4S$ (anal. Li, S), mp 170° ⟨34-1⟩.

Magnesium salt – $(C_{28}H_{43}O_4S)_2Mg \cdot 8H_2O$ (anal. Mg, S), mp 145-148° ⟨41-1⟩.

Potassium salt – $C_{28}H_{43}KO_4S$ (anal. K, S), mp 225° ⟨34-1⟩ • $C_{28}H_{43}KO_4S \cdot H_2O$ (anal. K, S), mp 211° ⟨41-1⟩ • mp 211° ⟨42-2⟩.

Pyridine salt – $C_{33}H_{49}NO_4S$ (anal. C_5H_5NH, S), mp 194-196° ⟨41-1⟩.

Sodium salt – $C_{28}H_{43}NaO_4S$ (anal. Na, S), mp 186° ⟨34-1⟩.

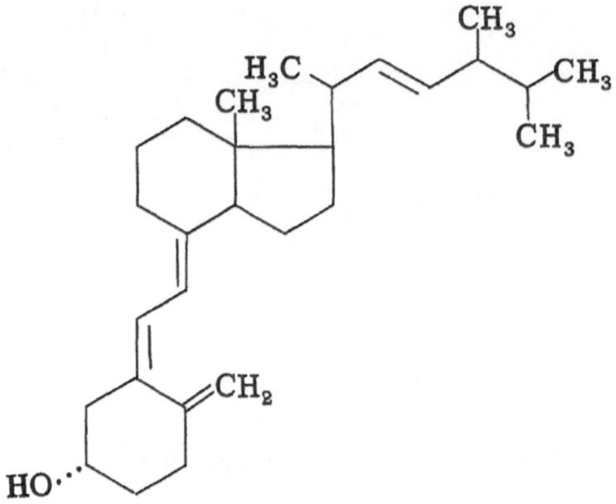

Vitamin D_2

3-SULFATE

Ammonium salt — $C_{28}H_{47}NO_4S$ (anal. C, H, S), mp 113-115°, $\lambda_{max}^{ethanol}$ 268 mμ ($E_{1cm}^{1\%}$ 572), IR, TLC ⟨65-10⟩ • mp 113-115°, λ_{max} 500 mμ ($E_{1cm}^{1\%}$ 1740) ⟨67-28⟩.

Sodium salt — $C_{28}H_{43}NaO_4S$ (anal. S), λ_{max} 500 mμ, TLC ⟨65-10⟩.

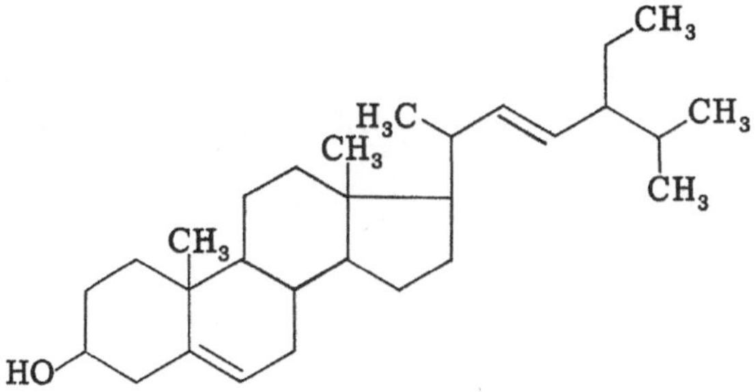

Stigmasta-5,22-dien-3β-ol
(Stigmasterol)

3-β-D-GLUCOPYRANOSIDURONIC ACID

2′,3′,4′-Tri-*O*-acetyl methyl ester — $C_{42}H_{66}O_{10}$ (anal. C, H, OCH_3), mp 172-174°, $[\alpha]_D^{21}$ − 52° (benzene) ⟨44-2⟩.

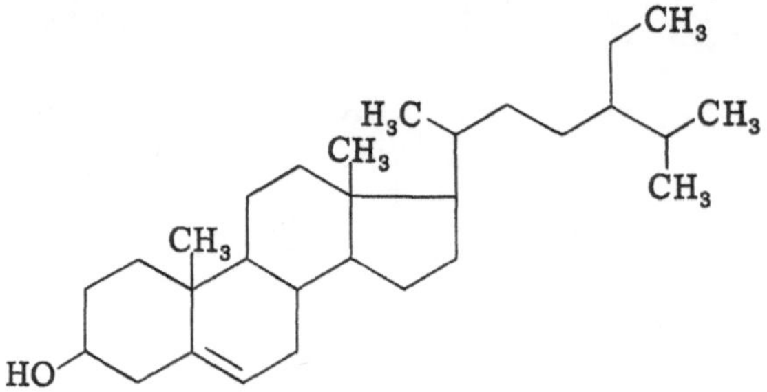

Stigmast-5-en-3β-ol
(β-Sitosterol)

3-β-D-GALACTOPYRANOSIDURONIC ACID

2′,3′,4′-Tri-*O*-acetyl methyl ester – $C_{42}H_{66}O_{10}$ (anal. C, H, OCH_3), mp 172-173°, $[\alpha]_D^{25}$ + 1.0° (chloroform) ⟨38-2⟩.

3-β-D-GLUCOPYRANOSIDURONIC ACID

Acid – $C_{35}H_{58}O_7$ (anal. C, H), mp 280-290° $[\alpha]_D^{20}$ – 30.2° (pyridine), ⟨59-9⟩ ⟨62-13⟩.

2′,3′,4′-Tri-*O*-acetyl – $C_{41}H_{64}O_{10}$ (anal. C, H), mp 163-164° IR ⟨59-9⟩ ⟨62-13⟩.

2′,3′,4′-Tri-*O*-acetyl methyl ester – $C_{42}H_{66}O_{10}$ (anal. C, H, OCH_3), mp 172-174°, $[\alpha]_D^{21}$ – 52° (benzene) ⟨44-2⟩.

3-SULFATE

Aluminum salt – mp 122-125° ⟨63-10⟩ • 5,6-dibromo ⟨63-10⟩.

Ammonium salt – mp 198-199° ⟨63-10⟩.

Calcium salt – mp 135-140° ⟨63-10⟩.

Lithium salt – mp 172-174° ⟨63-10⟩.

Methyl ester – $C_{30}H_{52}O_4S$ (anal. S), mp 105-106° ⟨61-9⟩.

Potassium salt – $C_{29}H_{49}KO_4S \cdot H_2O$ (anal. K), mp 216° ⟨61-9⟩ • 5,6-dibromo, mp 105-106° ⟨63-10⟩.

Pyridine salt – $C_{34}H_{55}NO_4S$ (anal. S), mp 184-185° ⟨61-9⟩ ⟨63-10⟩ • 5,6-dibromo – $C_{34}H_{55}Br_2NO_4S$ (anal. S), 134-135° ⟨61-9⟩.

Sodium salt – $C_{29}H_{49}NaO_4S \cdot 2H_2O$, mp 151-152° ⟨63-10⟩ • 5,6-dibromo, mp 102° ⟨63-10⟩.

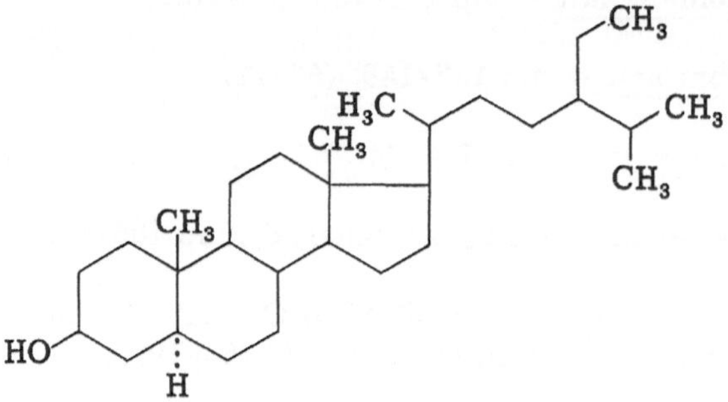

5 α-Stigmastan-3 β-ol

3-SULFATE

Pyridine salt – $C_{34}H_{57}NO_4S$ (anal. S), mp 188-190° ⟨61-9⟩.

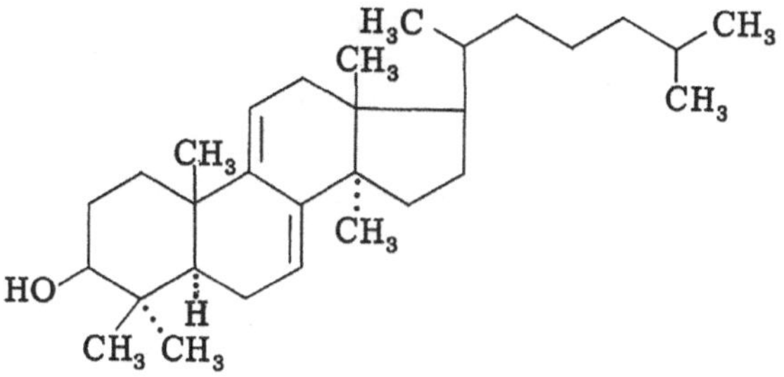

Lanosta-7,9(11)-dien-3β-ol

3-SULFATE

Pyridine salt – $C_{35}H_{55}NO_4S$ (anal. S), mp 234-235°, $[\alpha]_D^{18}$ + 57.3° (chloroform) ⟨52-1⟩.

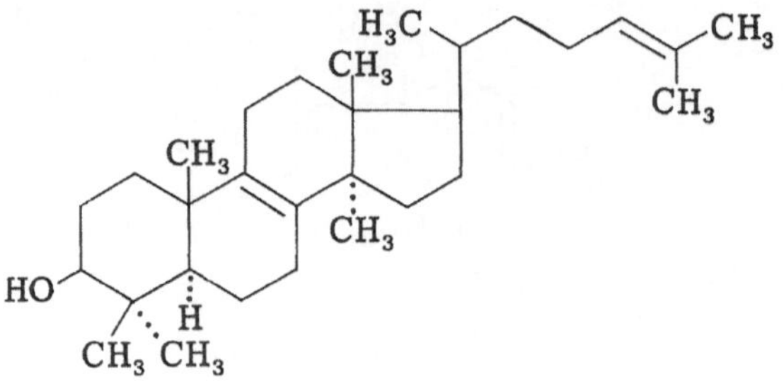

Lanosta-8,24-dien-3β-ol
(Lanosterol)

3-SULFATE

Potassium salt — $C_{30}H_{49}KO_4S$ (anal. K, S); mp 199-200° ⟨41-1⟩.

Pyridine salt — $C_{35}H_{55}NO_4S$ (anal. C_5H_5NH, S), mp 160-168° ⟨41-1⟩.

Allied Characterization Data

SULFATE

CC ⟨66-5⟩.

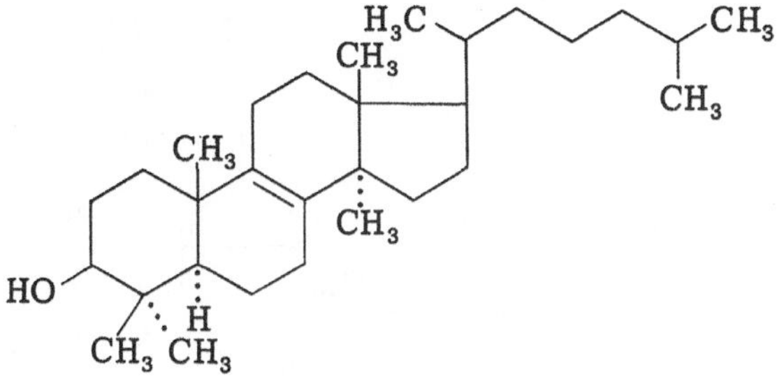

Lanost-8-en-3β-ol

3-SULFATE

Pyridine salt – $C_{35}H_{57}NO_4S$ (anal. C, H, S), mp 216-217°, $[\alpha]_D^{18}$ + 41.7° (chloroform) ⟨52-1⟩.

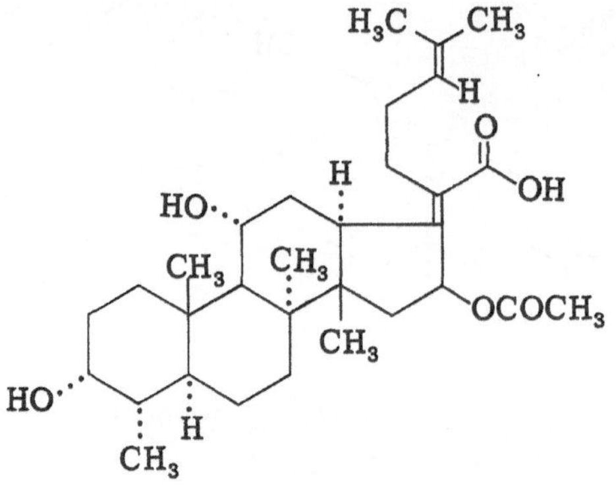

Fusidic acid

21-β-D-GLUCOPYRANOSIDURONIC ACID

$\underline{C_{21} \rightarrow C_{1'}\text{ ester}}$ — $C_{37}H_{56}O_{12}\cdot 1\frac{1}{2}H_2O$ (anal. C, H, H_2O), mp 158-162°, $[\alpha]_D^{20}$ − 32.3° (ethanol), λ_{max} 230 mμ (ε 9,100), IR, pK = 4.40, TLC ⟨66-18⟩.

References

15-1 Mandel, J. A. and Neuberg, C., *Biochem, Z.*, **71**, 186 (1915).

34-1 Natelson, S., Sobel, A. E. and Kramer, B., *J. Biol. Chem.*, **105**, 761 (1934).

35-1 Cohen, S. L. and Marrian, G. F., *J. Soc. Chem. Ind.*, **54**, 1025 (1935).

36-1 Cohen, S. L. and Marrian, G. F., *Biochem. J.*, **30**, 57 (1936).

36-2 Cohen, S. L., Marrian, G. F. and Odell, A. D., *Biochem. J.*, **30**, 2250 (1936).

36-3 Venning, E. M. and Browne, J. S. L., *Proc. Soc. Exptl. Biol. Med.*, **34**, 792 (1936).

38-1 Schachter, B. and Marrian, G. F., *J. Biol. Chem.*, **126**, 663 (1938).

38-2 Sell, H. M. and Link, K. P., *J. Biol. Chem.*, **125**, 235 (1938).

38-3 Schapiro, E., *Nature*, **142**, 1036 (1938).

39-1 Butenandt, A. and Hofstetter, H., *Z. Physiol. Chem.*, **259**, 222 (1939).

39-2 Natelson, S. and Gottfried, S. P., *J. Am. Chem. Soc.*, **61**, 971 (1939).

39-3 Schapiro, E., *Biochem. J.*, **33**, 385 (1939).

41-1 Sobel, A. E. and Spoerri, P. E., *J. Am. Chem. Soc.*, **63**, 1259 (1941).

42-1 Heckel, G. P., *N. Y. State J. Med.*, **42**, 2103 (1942).

42-2 Sobel, A. E. and Spoerri, P. E., *J. Am. Chem. Soc.*, **64**, 361 (1942).

42-3 Venning, E. H., Hoffman, M. M. and Browne, J. S. L., *J. Biol. Chem.*, **146**, 369 (1942).

42-4 Westphal, U., *Z. Physiol. Chem.*, **273**, 1 (1942).

42-5 Westphal, U., *Z. Physiol. Chem.*, **273**, 13 (1942).

43-1 Talbot, N. B., Ryan, J. and Wolfe, J. K., *J. Biol. Chem.*, **148**, 593 (1943).

44-1 Heard, R. D. H., Hoffman, M. M. and Mack, G. E., *J. Biol. Chem.*, **155**, 607 (1944).
44-2 Huebner, C. F., Overman, R. S. and Link, K. P., *J. Biol. Chem.*, **155**, 615 (1944).
44-3 Westphal, U., *Z. Physiol. Chem.*, 281, 14 (1944).

45-1 Mason, H. L. and Kepler, E. J., *J. Biol. Chem.*, **161**, 235 (1945).

47-1 Mason, H. L. and Strickler, H. S., *J. Biol. Chem.*, **171**, 543 (1947).
47-2 Sutherland, E. S. and Marrian, G. F., *Biochem. J.*,**41**, 193 (1947).

48-1 Klyne, W., Schachter, B. and Marrian, G. F., *Biochem. J.*, **43**, 231 (1948).
48-2 Lieberman, S., Dobriner, K., Hill, B. R., Fieser, L. F. and Rhoads, C. P., *J. Biol. Chem.*, **172**, 263 (1948).
48-3 Lieberman, S., Hariton, L. B. and Fukushima, D. K., *J. Am. Chem. Soc.*, 70, 1427 (1948).
48-4 Paterson, J. Y. F. and Klyne, W., *Biochem. J.*,**42**, 2P (1948).
48-5 Paterson, J. Y. F. and Klyne, W., *Biochem. J.*,**43**, 614 (1948).

49-1 Grant, G. A. and Glen, W. L., *J. Am. Chem. Soc.*, 71, 2255 (1949).
49-2 Holden, G. W., Levi, I. and Bromley, R., *J. Am. Chem. Soc.*, **71**, 3844 (1949).

50-1 Cohen, S. L., *J. Biol. Chem.*, **184**, 417 (1950).
50-2 Grant, G. A. and Beall, D., *Recent Progr. Hormone Res.*, **5**, 307 (1950).
50-3 Grant, G. A., Glen, W. L. and Barber, R. J., U.S. 2,534,121, Dec. 12, 1950.
50-4 Grant, J. K. and Marrian, G. F., *Biochem. J.*, **47**, 1 (1950).
50-5 Holden, G. W. and Bromley, R., *J. Am. Chem. Soc.*, **72**, 3807 (1950).

52-1 Birchenough, M. J. and Burton, H., *J. Chem. Soc.*, 2443 (1952).
52-2 Grant, G. A. and Glen, W. L., U.S. 2,597,723, May 20, 1952. Brit. 663,321, Dec. 1, 1951.

53-1 Hasbrouck, R. B., U.S. 2,642,427, June 16, 1953.
53-2 Salkin, R., U.S. 2,636,042, April 21, 1953.
53-3 Teich, S., Rogers, J., Lieberman, S., Engel, L. L. and Davis, J. W., *J. Am. Chem. Soc.*, **75**, 2523 (1953).

54-1 Apotheker, D., Owades, J. L. and Sobel, A. E., *J. Am. Chem. Soc.*, **76**, 3684 (1954).
54-2 Hasbrouck, R. B., U.S. 2,666,066, Jan. 12, 1954.

55-1 Fukushima, D. K., Leeds, N. S., Bradlow, H. L., Kritchevsky, T. H., Stokem, M. B. and Gallagher, T. F., *J. Biol. Chem.*, **212**, 449 (1955).
55-2 Lewbart, M. L. and Schneider, J. J., *Nature*, **176**,1175 (1955).
55-3 Schneider, J. J., Lewbart, M. L., Levitan, P. and Lieberman, S., *J. Am. Chem. Soc.*, **77**, 4184 (1955).

56-1 D'Alo, F., *Farmaco (Pavia) Ed. Sci.*, **11**, 477 (1956).
56-2 D'Alo, F., *Farmaco (Pavia) Ed. Sci.*, **11**, 1004 (1956).
56-3 Isselbacher, K. J., *Recent Progr. Hormone Res.*, **2**, 134 (1956).
56-4 Lewbart, M. L. and Schneider, J. J., *Federation Proc.*, **15**, 300 (1956).
56-5 Schneider, J. J. and Lewbart, M. L., *J. Biol. Chem.*, **222**, 787 (1956).

57-1 Bush, I. E., *Biochem. J.*, **67**, 23P (1957).
57-2 Cavina, G., *Rend. Ist. Super. Sanita*, **20**, 923 (1957).
57-3 Cavina, G. and Tentori, L., *Rend. Ist. Super. Sanita*, **20**, 951 (1957).
57-4 Crepy, O., Jayle, M. F. and Meslin, F., *Compt. Rend. Soc. Biol.*, **151**, 234 (1957).
57-5 D'Alo, F., *Farmaco (Pavia), Ed. Sci.*,**12**, 166 (1957).
57-6 Dodgson, K. S. and Spencer, B., *Methods Biochem. Anal.*,**4**, 211 (1957).
57-7 McKenna, J. and Norymberski, J. K., *J. Chem. Soc.*, 3889 (1957).
57-8 McKenna, J. and Norymberski, J. K., *J. Chem. Soc.*, 3893 (1957).
57-9 Patel, D. K., Petrow, V. and Stuart-Webb, I. A., *J. Chem. Soc.*, 665 (1957).
57-10 Sie, H. and Fishman, W. H., *J. Biol. Chem.*, **225**, 453 (1957).

57-11 Wotiz, H. H., Leftin, J. H., Smakula, E. and Lichtin, N. N., Abstr. 128, 131st. Meeting of the American Chemical Society, Miami Beach, Fla., Apr. 1957, p. 58c.

58-1 Burstein, S. and Lieberman, S., *J. Biol. Chem.*, **233**, 331 (1958).
58-2 Burstein, S. and Lieberman, S., *J. Am. Chem. Soc.*, 80, 5235 (1958).
58-3 Cavina, G. and Tentori, L., *Clin. Chim. Acta*, **3**, 160 (1958).
58-4 DeMeio, R. H., Lewycka, C., Wizerkaniuk, M. and Salciunas, O., *Biochem. J.*, **68**, 1 (1958).
58-5 Griebsch, E. and Garn, W., Ger. 1,047,780, Dec. 31, 1958; *Chem. Abstr.*, 55, 3661b (1961).
58-6 Griebsch, E., Zaehlsdorff, G. and Pirner, K., U. S. 2,828,306, Mar. 25, 1958.
58-7 Hadd, H. E., Abstr. 131, 134th. Meeting of the American Chemical Society, Chicago, Ill., Sept. 1958, p. 54c.
58-8 Pelzer, H., Staib, W. and Ott, D., *Z. Physiol. Chem.*, **312**, 15 (1958).
58-9 Wotiz, H. H., Sie, H. and Fishman, W. H., *J. Biol. Chem.*, **232**, 723 (1958).
58-10 Zorbach, W. W., *J. Org. Chem.*, **23**, 1797 (1958).

59-1 Cohn, G. L. and Bondy, P. K. *J. Biol. Chem.*, **234**, 31 (1959).
59-2 Griebsch, E. and Garn, W., Ger. 1,066,581, Oct. 8, 1959; *Chem. Abstr.*, **55**, 9477g (1961).
59-3 Pelzer, H., *Z. Physiol. Chem.*, **314**, 234 (1959).
59-4 Purdy, R. H., Engel, L. L., and Oncley, J. L., *Federation Proc.*, 18, 305 (1959).
59-5 Schmid, O. J., Oriol-Bosch, A. and Voigt, K. D., *Clin. Chim. Acta*, **4**, 599 (1959).
59-6 Schneider, J. J. and Lewbart, M. L., *Recent Progr. Hormone Res.*, **15**, 201 (1959).
59-7 Smakula, E., Leftin, J. H. and Wotiz, H. H., *J. Am. Chem. Soc.*, **81**, 1704 (1959).
59-8 Smakula, E., Lichtin, N. N., Leftin, J. H. and Wotiz, H. H., *J. Am. Chem. Soc.*, **81**, 1708 (1959).
59-9 Vrkoč, J., Herout, V. and Šorm, F., *Collect. Czech. Chem. Commun.*, **24**, 3938 (1959).

60-1 Baulieu, E. E., *Compt. Rend.*, **250**, 4219 (1960).

60-2 Baulieu, E. E., *Compt. Rend.*, **251**, 1421 (1960).

60-3 Baulieu, E. E. and Emiliozzi, R., *Compt. Rend.*, 251, 3106 (1960).

60-4 Brooksbank, B. W. L. and Haslewood, G. A. D., *Acta Endocrinol.*, *Suppl.* **51**, 1023 (1960).

60-5 Burstein, S., Jacobsohn, G. M. and Lieberman, S., *J. Am. Chem. Soc.*, **82**, 1226 (1960).

60-6 Emiliozzi, R., *Bull. Soc. Chim. France*, 911 (1960).

60-7 Feinstone, W. H., U.S. 2,928,848, Mar. 15, 1960.

60-8 Foggitt, F. and Kellie, A. E., *Biochem. J.*, **76**, 62P (1960).

60-9 Griebsch, E. and Garn, W., Ger. 1,090,208, Oct. 6, 1960; *Chem. Abstr.*, **55**, 26044b (1961).

60-10 Hadd, H. E. and Dorfman, R. I., Abstr. 12, 137th Meeting of the American Chemical Society, Cleveland, Ohio, Apr. 1960, p. 6c.

60-11 Levitz, M., Condon, G. P., Money, M. L. and Dancis, J., *J. Biol. Chem.*, **235**, 973 (1960).

60-12 Lucas, R. A., Dickel, D. F., Dziemian, R. L., Ceglowski, M. J., Hensle, B. L. and MacPhillamy, H. B., *J. Am. Chem. Soc.*, **82**, 5688 (1960).

60-13 Mason, M. and Gullekson, E., *J. Biol. Chem.*, **235**, 1312 (1960).

60-14 Pasqualini, J., *Compt. Rend.*, **250**, 3892 (1960).

60-15 Pasqualini, J., Zelnik, R. and Jayle, M. F., *Experientia*, **16**, 317 (1960).

60-16 Sheps, M. C., Purdy, R. H., Engel, L. L. and Oncley, J. L., *J. Biol. Chem.*, **235**, 3042 (1960).

60-17 Staib, W. and Dönges, K., *Z. Physiol. Chem.*, **319**, 233 (1960).

60-18 Zelnik, R., Desfosses, B. and Emiliozzi, R., *Compt. Rend.*, **250**, 1671 (1960).

61-1 Baulieu, E. E., Emiliozzi, R. and Corpéchot., C., *Experientia*, **17**, 110 (1961).

61-2 Boström, H., *Acta Endocrinol.*, **37**, 405 (1961).

61-3 Brooksbank, B. W. L. and Haslewood, G. A. D., *Biochem. J.*, **80**, 488 (1961).

61-4 Carpenter, J. G. D., *Dissertation*, University of London, May 1961.

61-5 Carpenter, J. G. D. and Kellie, A. E., *Biochem. J.*, **78**, 1P (1961).

61-6 Conrad, S., Mahesh, V. and Herrmann, W., *J. Clin. Invest.*, 40, 947 (1961).

61-7 Diczfalusy, E., Cassmer, O., Alonso, C. and deMiquel, M., *Acta Endocrinol.*, **38**, 31 (1961).
61-8 Felger, C. B. and Katzman, P. A., *Federation Proc.*, **20**, 199 (1961).
61-9 Khaletskii, A. M. and Vasil'eva, M. V., *Zh. Obshch. Khim.*, **31**, 2996 (1961).
61-10 Menini, E. and Diczfalusy, E., *Endocrinology*, **68**, 492 (1961).
61-11 Neeman, M. and Hashimoto, Y., *Tetrahedron Lett.*, 183 (1961).
61-12 Oertel, G. W., *Naturwissenschaften*, **48**, 621 (1961).
61-13 Pasqualini, J. and Jayle, M. F., *Biochem. J.*, **81**, 147 (1961).
61-14 Pasqualini, J. R. and Jayle, M. F., *Experientia*, **17**, 450 (1961).
61-15 Troen, P., Nilsson, B., Wiqvist, N. and Diczfalusy, E., *Acta Endocrinol.*, **38**, 361 (1961).
61-16 Uete, T., *Gifu Ika Daigaku Kiyo*, **9**, 721 (1961).
61-17 Winguth, K., *Dissertation*, University of Hamburg, 1961.

62-1 Baulieu, E. E., *J. Clin. Endocrinol. Metab.*, **22**, 501 (1962).
62-2 Baulieu, E. E. and Emiliozzi, R., *Bull. Soc. Chim. Biol.*, **44**, 823 (1962).
62-3 Carpenter, J. G. D. and Kellie, A. E., *Biochem. J.*, **84**, 303 (1962).
62-4 Diczfalusy, E., Franksson, C., Lisboa, B. P. and Martinsen, B., *Acta Endocrinol.*, **40**, 537 (1962).
62-5 Levitz, M., Emerman, s., Dancis, J., Tillinger, K. G., Wiqvist, N. and Diczfalusy, E., *Federation Proc.*, **21**, 212 (1962).
62-6 Neeman, M. and Hashimoto, Y., *J. Am. Chem. Soc.*, **84**, 2972 (1962).
62-7 Nimke, G., Ger. 1,122,520, Jan. 25, 1962; *Chem. Abstr.*, **57**, 7354c (1962).
62-8 Oertel, G. W. and Kaiser, E., *Biochem. Z.*, **336**, 10 (1962).
62-9 Pasqualini, J. R. and Jayle, M. F., *J. Clin. Invest.*, **41**, 981 (1962).
62-10 Pasqualini, J. R., Zelnik, R. and Jayle, M. F., *Bull. Soc. Chim. France*, 1171 (1962).
62-11 Shindo, M., Japan 62/4527, June 14, 1962; *Chem. Abstr.*, **58**, 11462d (1963).

62-12 Starka, L., Sulcova, J. and Silink, K., *Clin. Chim. Acta*, 7, 309 (1962).
62-13 Vrkoč, J., *Collect. Czech. Chem. Commun.*, 27, 1345 (1962).
62-14 Wengle, B. and Boström, H., *Acta Chem. Scand.*, 16, 502 (1962).
62-15 Wotiz, H. H., *Biochim. Biophys. Acta*, 60, 28 (1962).
62-16 Zorbach, W. W., Valiaveedan, G. D. and Kashelikar, D. V., *J. Org. Chem.*, 27, 1766 (1962).

63-1 Baulieu, E. E., Corpéchot, C. and Emiliozzi, R., *Steroids*, **2**, 429 (1963).
63-2 Beling, C. G., *Acta Endocrinol., Suppl.* 79, (1963).
63-3 Calvin, H. I., Vande Wiele, R. L. and Lieberman, S., *Biochemistry*, 2, 648 (1963).
63-4 Clark, A. F. and Solomon, S., *J. Clin. Endocrinol. Metab.*, **23**, 481 (1963).
63-5 Cohn, G. L., Mulrow, P. J. and Dunne, V. C., *J. Clin. Endocrinol. Metab.*, 23, 671 (1963).
63-6 Engelfried, O., Schargan, K. and Wiechert, R., Ger. 1,152,105, Aug. 1, 1963; Der. 6589.
63-7 Hadd, H. E. and Dorfman, R. I., *J. Biol. Chem.*, **238**, 907 (1963).
63-8 Hashimoto, Y. and Neeman, M., *J. Biol. Chem.*, **238**, 1273 (1963).
63-9 Hirschmann, H. and Williams, J. S., *J. Biol. Chem.*, **238**, 2305 (1963).
63-10 Khaletskii, A. M. and Vasil'eva, M. V., *Zh. Obshch. Khim.*, **33**, 1104 (1963).
63-11 Pasqualini, J. R., Dutter, F. and Jayle, M. F., *Biochim. Biophys. Acta*, **69**, 331 (1963).
63-12 Pasqualini, J. R. and Jayle, M. F., *Compt. Rend.*, 257, 2345 (1963).
63-13 Pasqualini, J. R., Uhrich, F. and Jayle, M. F., *Bull. Soc. Chim. Biol.*, **45**, 695 (1963).
63-14 Payne, A. H. and Mason, M., *Biochim. Biophys. Acta*, 71, 719 (1963).
63-15 Sarett, L. H., Strachan, R. G. and Hirschmann, R., Belg. 623,144, April 3, 1963; Der. 7162.
63-16 Siiteri, P. K., *Dissertation*, Columbia Univ., 1963.
63-17 Siiteri, P. K., Vande Wiele, R. L. and Lieberman, S., *J. Clin. Endocrinol. Metab.*, **23**, 588 (1963).
63-18 Wallace, E. Z. and Lieberman, S., *J. Clin. Endocrinol. Metab.*, **23**, 90 (1963).

63-19 Wengle, B. and Boström, H., *Acta Chem. Scand.*, **17**, 1203 (1963).

63-20 Wotiz, H. H. and Fishman, W. H., *Steroids*, **1**, 211 (1963).

64-1 Boström, H. and Wengle, B., *Acta Soc. Med. Upsalien.*, **69**, 41 (1964).

64-2 Calvin, H. I. and Lieberman, S., *Biochemistry*, **3**, 259 (1964).

64-3 Crépy, O., Judas, O. and Lachese, B., *J. Chromatogr.*, **16**, 340 (1964).

64-4 Elce, J. S., Carpenter, J. G. D. and Kellie, A. E., *Biochem. J.*, **91**, 30P (1964).

64-5 Emiliozzi, R., *Compt. Rend.*, **258**, 3875 (1964).

64-6 Foggitt, F. and Kellie, A. E., *Biochem. J.*, **91**, 209 (1964).

64-7 Griebsch, E. and Garn, W., U.S. 3,152,044, Oct. 6, 1964.

64-8 Hadd, H. E. and Blickenstaff, R. T., *Steroids*, **4**, 503 (1964).

64-9 Hirschmann, R., Strachan, R. G., Buchschacher, P., Sarett, L. H., Steelman, S. L. and Silber, R., *J. Am. Chem. Soc.*, **86**, 3903 (1964).

64-10 Kellie, A. E., *Structure and Metabolism of Corticosteroids*, J. Pasqualini and M. Jayle, Ed., Academic Press, New York, N. Y. 1964, p 21.

64-11 Kornel, L., *J. Clin. Endocrinol. Metab.*, **24**, 956 (1964).

64-12 Kornel, L., Kleber, J. W. and Conine, J. W., *Steroids*, **4**, 67 (1964).

64-13 Layne, D. S., Sheth, N. A. and Kirdani, R. Y., *J. Biol. Chem.*, **239**, 3221 (1964).

64-14 Nitta, Y., Shindo, M. and Takamura, K., *Chem. Pharm. Bull. (Tokyo)*, **12**, 450, (1964).

64-15 Pasqualini, J. R., *Aldosterone Symp. Prague* 1963, p. 131 (pub. 1964).

64-16 Pasqualini, J. R., *Arch. Biochem. Biophys.*, 106, 15 (1964).

64-17 Pasqualini, J. R., *Compt. Rend.*, **259**, 934 (1964).

64-18 Pasqualini, J. R. and Faggett, J., *J. Endocrinol.*, **31**, 85 (1964).

64-19 Rhone-Poulenc S. A., Fr. M2624, July 24, 1964; Der. 12462.

64-20 Robel, P., Emiliozzi, R. and Baulieu, E. E., *Compt. Rend.*, **258**, 1331 (1964).

64-21 Roberts, K. D., Bandi, L., Calvin, H. I., Drucker, W. D. and Lieberman, S., *Biochemistry*, **3**, 1983 (1964).

64-22 Roberts, K. D. Bandi, L., Calvin, H. I., Drucker, W. D. and Lieberman, S., *J. Am. Chem. Soc.*, 86, 958 (1964).

64-23 Schering AG., Ger. 1,170,949, May 27, 1964; Der. 12236.
64-24 Schneider, J. J., *Hormonal Steroids*, Proc. Intern. Congr. Hormonal Steroids, 1st., Milan, 1962, p. 127 (pub. 1964).
64-25 Slaunwhite, W. R. Jr., Lichtman, M. A. and Sandberg, A. A., *J. Clin. Endocrinol. Metab.*, **24**, 638 (1964).
64-26 Wilson, R., Eriksson, G. and Diczfalusy, E., *Acta Endocrinol.*, **46**, 525 (1964).
64-27 Zorbach, W. W. and Valiaveedan, G. D., *J. Org. Chem.*, **29**, 2462 (1964).

65-1 Baulieu, E. E. and Corpéchot, C., *Bull. Soc. Chim. Biol.*, **47**, 443 (1965).
65-2 Becker, J. F., *Biochim. Biophys. Acta*, 100, 574 (1965).
65-3 Bird, C. E., Solomon, S., Wiqvist, N. and Diczfalusy, E., *Biochim. Biophys. Acta*, **104**, 623 (1965).
65-4 Cocker, J. D., Elks, J., May, P. J., Nice, F. A., Phillipps, G. H., and Wall, W. F., *J. Med. Chem.*, 8, 417 (1965).
65-5 Crépy, O. and Jayle, M. F., *Bull. Soc. Chim. Biol.*, **47**, 427 (1965).
65-6 Deghenghi, R. and Revesz, C., *J. Endocrinol.*, 31, 301 (1965).
65-7 Elce, J. S., *Dissertation*, University of London, 1965.
65-8 Fishman, W. H., Harris, F. and Green, S., *Steroids*, 5, 375 (1965).
65-9 Hähnel, R., *Anal. Biochem.*, 10, 184 (1965).
65-10 Higaki, M., Takahashi, M., Suzoki, T. and Sahashi, Y., *J. Vitaminol. (Kyoto)*, 11, 261 (1965).
65-11 Jayle, M. F., *Analysis of Steroid Hormones*, Vol. III, Infrared Spectra in 1300-450 cm^{-1} Region, Masson, Paris, 1965.
65-12 Jirku, H. and Layne, D. S., *Biochemistry*, **4**, 2126 (1965).
65-13 Kirdani, R. Y., *Steroids*, 6, 845 (1965).
65-14 Klein, G. P. and Giroud, C. J. P., *Steroids*, **5**, 765 (1965).
65-15 Layne, D. S., *Endocrinology*, **76**, 600 (1965).
65-16 Levitz, M., Katz, J. and Twombly, G. H., *Steroids*, 6, 553 (1965).
65-17 Levitz, M., Katz, J. and Twombly, G. H., *Federation Proc.*, **24**, 534 (1965).
65-18 Mattox, V. R., Vrieze, W. and Goodrich, J., *Federation Proc.*, 24, 415 (1965).
65-19 Neudert, W. and Röpke, H., *Atlas of Steroid Spectra*, Springer-Verlag, New York Inc., 1965.

65-20 Pasqualini, J. R., *Bull. Soc. Chim. Biol.*,**47**, 471 (1965).
65-21 Pasqualini, J. R., Uhrich, F. and Jayle, M. F., *Biochim. Biophys. Acta,* **104**, 515 (1965).
65-22 Payne, A. H. and Mason, M., *Steroids,* **5**, 21 (1965).
65-23 Robel, P., Emiliozzi, R. and Baulieu, E. E., *Compt. Rend.*, **261**, 4886 (1965).
65-24 Schriefers, H., Kley, H. K. and Otto, M., *Z. Physiol. Chem.*, **341**, 215 (1965).
65-25 Schwers, J., Eriksson, G., Wiqvist, N. and Diczfalusy, E., *Biochim. Biophys. Acta,* 100, 313 (1965).
65-26 Villax, I., Ger. 1,200,822, Sept. 16, 1965; Der. 5880.
65-27 Williams, K. I. H., Smulowitz, M. and Fukushima, D. K., *J. Org. Chem.*, **30**, 1447 (1965).

66-1 Ayerst, McKenna & Harrison, Ltd., Neth. 6,509,045, Jan. 17, 1966; Der. 19901.
66-2 Breuer, H. and Wessendorf, D., *Z. Physiol. Chem.*, **345**, 1 (1966).
66-3 Calvin, H. I., *Dissertation Abstr.*, **26**, 5690 (1966) (Columbia Univ. 1965).
66-4 Calvin, H. I. and Lieberman, S., *J. Clin. Endocrinol. Metab.*, **26**, 402 (1966).
66-5 Calvin, H. I., Roberts, K. D., Weiss, C., Bandi, L., Cos, J. J. and Lieberman, S., *Anal. Biochem.*, **15**, 426 (1966).
66-6 Cantrall, E. W., McGrath, M. G. and Bernstein, S., *Steroids,* **8**, 967 (1966).
66-7 Cantrall, E. W., Littell, R. and Bernstein, S., U.S. 3,274,056, Sept. 20, 1966.
66-8 Corbellini, A., Torti, G., Boffi, C. and Ferrara, G., *Rend. Ist. Lombardo Sci. Lettere, A.*, **100**, 273 (1966).
66-9 Crépy, O., Judas, O. and Jayle, M. F., *Compt. Rend. Soc. Biol.*, **160**, 891 (1966).
66-10 Crépy, O., Pasqualini, J. R., Ducret, M. A. and Jayle, M. F., *European J. Steroids,* **1**, 195 (1966).
66-11 Dahm, K. and Breuer, H., *Z. Klin. Chem.*,**4**, 153 (1966).
66-12 Dahm, K. and Breuer, H., *Biochim. Biophys. Acta,* **128**, 306 (1966).
66-13 Dahm, K. and Breuer, H., *Acta Endocrinol.*, **52**, 43 (1966).
66-14 Dean, P. D. G. and Whitehouse, M. W., *Biochem. J.*, **98**, 410 (1966).
66-15 Dray, M. F. and Ledru, M., *Compt. Rend.*,**262**, 679 (1966).

66-16 Easterling, W. E. Jr., Simmer, H. H., Dignam, W. J., Frankland, M. V. and Naftolin, F., *Steroids*, **8**, 157 (1966).

66-17 Ferrara, G., Boffi, C., Torti, G. and Corbellini, A., *Steroids*, **8**, 111 (1966).

66-18 Godtfredsen, W. O. and Vangedal, S., *Acta Chem. Scand.*, **20**, 1599 (1966).

66-19 Goebelsmann, U., Cooke, I., Wiqvist, N. and Diczfalusy, E., *Acta Endocrinol.*, **52**, 30 (1966).

66-20 Goebelsmann, U., Erigksson, G., Diczfalusy, E., Levitz, M. and Condon, G. P., *Acta Endocrinol.*, **53**, 391 (1966).

66-21 Griffiths, K., Grant, J. K., Browning, M. C. K., Cunningham, D. and Barr, G., *J. Endocrinol.*, **35**, 299 (1966).

66-22 Hadd, H. E., *Dissertation Abstr.*, **26**, 5009 (1966) (Indiana Univ., 1964).

66-23 Hähnel, R. and Rahman, M. G. B. A., *Clin. Chim. Acta*, **13**, 797 (1966).

66-24 Haslewood, G. A. D., *Biochem. J.*, 100, 233 (1966).

66-25 Joseph, J. P., Dusza, J. P. and Bernstein, S., *Steroids*, **7**, 577 (1966).

66-26 Kielmann, N., Stachenko, J. and Giroud, C. J. P., *Steroids*, **8**, 993 (1966).

66-27 Kirschner, M. A., Wiqvist, N. and Diczfalusy, E., *Acta Endocrinol.*, **53**, 584 (1966).

66-28 Klein, G. P. and Giroud, C. J. P., *Can. J. Biochem.*, **44**, 1005 (1966).

66-29 Morita, K., Ger. 1,224,738, Sept. 15, 1966; Der. 22851; Japan 63/14731; Der. 8867.

66-30 Mumma, R. O., *Lipids*, **1**, 221 (1966).

66-31 Nagayama, F., Saito, A. and Idler, D. R., *Can. J. Biochem.*, **44**, 1109 (1966).

66-32 Oertel, G. W. and Knapstein, P., *Z. Physiol. Chem.*, **344**, 159 (1966).

66-33 Ozon, R. and Breuer, H., *Gen. Compt. Endocrinol.*, **6**, 295 (1966).

66-34 Pasqualini, J. R., Dutter, F. and Jayle, M. F., *J. Endocrinol.*, **35**, 145 (1966).

66-35 Robel, P., Emiliozzi, R. and Baulieu, E. E., *J. Biol. Chem.*, **241**, 20 (1966).

66-36 Robel, P., Emiliozzi, R. and Baulieu, E. E. *J. Biol. Chem.*, **241**, 5879 (1966).

66-37 Sarfaty, G. A. and Lipsett, M. B., *Anal. Biochem.*, **15**, 184 (1966).

66-38 Schriefers, H., Ghraf, R. and Pohl, F., *Z. Physiol. Chem.*, **344**, 25 (1966).

66-39 Segaloff, A., Gabbard, R. B. and Carriere, B. T., *Steroids*, **7**, 137 (1966).
66-40 Sjovall, K., Sjovall, J., Maddock, K. and Horning, E. C., *Anal. Biochem.*, **14**, 337 (1966).
66-41 Sjovall, J. and Vihko, R., *Acta Chem. Scand.*, **20**, 1419 (1966).
66-42 Starka, L., Janata, J., Breuer, H. and Hampl, R., *European J. Steroids*, **1**, 37 (1966).
66-43 Tockstein, G. V., *Dissertation*, St. Louis Univ., 1966.
66-44 Troen, P., deMiquel, M. and Alonso, C., *Biochemistry*, **5**, 332 (1966).

67-1 Anderson, I. G., Haslewood, G. A. D., Cross, A. D. and Tokes, L., *Biochem. J.*, **104**, 1061 (1967).
67-2 Arcos, M. and Lieberman, S., *Biochemistry*, **6**, 2032 (1967).
67-3 Collins, D. C., Williams, K. I. H. and Layne, D. S., *Arch. Biochem. Biophys.*, **121**, 609 (1967).
67-4 Collins, D. C., Williams, K. I. H. and Layne, D. S., *Endocrinology*, **80**, 893 (1967).
67-5 Conrad, S. H., Pion, R. J. and Kitchin, J. D. III, *J. Clin. Endocrinol. Metab.*, **27**, 114 (1967).
67-6 Coulon-Morelec, M. J., *Bull. Soc. Chim. Biol.*, **49**, 825 (1967).
67-7 Creange, J. E. and Szego, C. M., *Biochem. J.*, **102**, 898 (1967).
67-8 Dray, F., Mowszowicz, I. and Ledru, M., *Steroids*, **10**, 501 (1967).
67-9 Drayer, N. and Lieberman, S., *J. Clin. Endocrinol. Metab.*, **27**, 136 (1967).
67-10 Elce, J. S., Carpenter, J. G. D. and Kellie, A. E., *J. Chem. Soc.* (C), 542 (1967).
67-11 Emerman, S. Twombly, G. H. and Levitz, M., *J. Clin. Endocrinol. Metab.*, **27**, 539 (1967).
67-12 Francis, F. E. and Kinsella, R. A. Jr., *J. Clin. Endocrinol. Metab.*, **27**, 211 (1967).
67-13 Hähnel, R., *J. Endocrinology*, **38**, 417 (1967).
67-14 Horning, E. C., Horning, M. G., Ikekawa, N., Chambaz, E. M., Jaakonmaki, P. I. and Brooks, C. J. W., *Gas Chromatogr.*, **5**, 283 (1967).
67-15 Hurwitz, A. R., Burke, H. J., Marra, R. A., Abstr. 18, Meeting American Pharmaceutical Association, Las Vegas, Nevada, April 1967.

67-16 Jaakomaki, P. I., Yarger, K. A. and Horning, E. C., *Biochem. Biophys. Acta,* **137**, 216 (1967).
67-17 Joseph, J. P., Dusza, J. P. and Bernstein, S.,*J. Am. Chem. Soc.,* **89**, 5078 (1967).
67-18 Klein, G. P. and Giroud, C. J. P.,*Steroids,* **9**, 113 (1967).
67-19 Knapstein, P., Rindt, W., Wendlberger, F. and Oertel, G. W.,*Z. Physiol. Chem.,* **348**, 93 (1967).
67-20 Mattox, V. R. and Vrieze, W.,*Federation Proc.,* **26**, 425 (1967).
67-21 Miklosi, S. A. and McCosker, P. J.,*J. Endocrinol.,* **39**, 361 (1967).
67-22 Nambara, T. and Imai, K., *Chem. Pharm. Bull. (Tokyo),* **15**, 1232 (1967).
67-23 Oertel, G. W., Trieber, L. and Rindt, W., *Experientia,* **23**, 97 (1967).
67-24 Pierrepoint, C. G.,*Anal. Biochem.,* **18**, 181 (1967).
67-25 Rahman, M. G. B. A. and Hähnel, R., *Clin. Chim. Acta,* **17**, 59 (1967).
67-26 Robel, P., Emiliozzi, R. and Baulieu, E. E., *J. Clin. Endocrinol. Metab.,* 27, 1290 (1967).
67-27 Sahashi, Y., Suzuki, T., Higaki, M. and Asano, T., *J. Vitaminol. (Kyoto),* **13**, 33 (1967).
67-28 Sahashi, Y., Suzuki, T., Higaki, M., Takahashi, M., Asano, T., Hasegawa, T., and Miyazawa, E., *J. Vitaminol. (Kyoto),* 13, 37 (1967).
67-29 Smith, E. R. and Kellie, A. E., *Biochem. J.,* **104**, 83 (1967).
67-30 Tamm, J., Volkwein, U. and Voigt, K. D.,*Experientia,* **23**, 299 (1967).
67-31 Vanden Heuvel, W. J. A., *J. Chromatogr.,* **28**, 406 (1967).
67-32 Wang, D. Y., Bulbrook, R. D., Ellis, F. and Coombs, M. M.,*J. Endocrinol.,* **39**, 395 (1967).
67-33 YoungLai, E., and Solomon, S., *Endocrinology,* **80**, 177 (1967).
67-34 YoungLai, E. and Solomon, S.,*Biochemistry,* **6**, 2040 (1967).
67-35 Zucconi, G., Goebelsmann, U., Wiqvist, N. and Diczfalusy, E.,*Acta Endocrinol.,* **56**, 71 (1967).

68-1 Cantrall, E. W., *unpublished.*
68-2 Dusza, J. P., Joseph, J. P. and Bernstein, S.,*Steroids,* **12** in press (1968).
68-3 Dusza, J. P., Joseph, J. P. and Bernstein, S., *unpublished.*

68-4 Fex, H., *private communication.*

68-5 Fukushima, D. and Sauer, G., *private communication.*

68-6 Heller, M., Lenhard, R. H. and Bernstein, S., *unpublished.*

68-7 Littell, R., Cantrall, E. W. and Bernstein, S., *unpublished.*

68-8 Miyazaki, M. and Fishman, J., *J. Org. Chem.*, **33**, 662 (1968).